Strategic Program for Innovations in Undergraduate Physics: Project Report

D0129469

edited by

Robert C. Hilborn
Amherst College

Ruth H. Howes
Ball State University

Kenneth S. Krane
Oregon State University

With Support from:

The ExxonMobil Foundation
American Association of Physics Teachers
American Institute of Physics
American Physical Society

January 2003

Strategic Programs for Innovations in Undergraduate Physics: Project Report

Published and Distributed by:

The American Association of Physics Teachers
One Physics Ellipse
College Park, MD 20740-3845
U.S.A.
www.aapt.org
301-209-3333

This project was supported by:

The ExxonMobil Foundation
American Association of Physics Teachers
American Institute of Physics
American Physical Society

Cover design by Kim Wolford

ISBN 1-931024-03-0

Table of Contents

Preface

This report describes the results of the project Strategic Programs for Innovations in Undergraduate Physics (SPIN-UP), organized by the National Task Force on Undergraduate Physics. The Task Force received support for SPIN-UP from the American Association of Physics Teachers, the American Physical Society, the American Institute of Physics, and a generous grant from the ExxonMobil Foundation. Particular thanks go to Edward F. Ahnert, President of the ExxonMobil Foundation, Truman T. Bell, program officer, and Jean Moon, consultant to the Foundation. Their assistance in shaping and focusing the goals of the project was invaluable. We gratefully acknowledge Roman Czujko and his colleagues at the American Institute of Physics Statistical Research Center for their work on the survey of all bachelor's degree-granting physics programs in the United States. Although led by the Task Force, SPIN-UP benefited from the volunteer efforts of more than 50 physicists for the site visits, the hospitality and work of the faculty of the 21 physics departments visited as part of the site visit program, and the generous time spent on the survey by 74% of the physics departments in the country. SPIN-UP is indeed a physics community effort.

Executive Summary

Strategic Programs for Innovations in Undergraduate Physics (SPIN-UP) set out to answer an intriguing question: Why, in the 1990s, did some physics departments increase the number of bachelor's degrees awarded in physics or maintain a number much higher than the national average for their type of institution? During that decade, the number of bachelor's degrees awarded in the physical sciences, engineering, and mathematics declined across the country. Yet in the midst of this decline some departments had thriving programs. What made these departments different? What lessons can be learned to help departments in the sciences, engineering, and mathematics that are—to put it generously—less than thriving? SPIN-UP, a project of the National Task Force on Undergraduate Physics, set out to answer these questions by sending site visit teams to 21 physics departments whose undergraduate programs were, by various measures, thriving. These visits took place mostly during the 2001–2002 academic year. In addition, with the aid of the American Institute of Physics Statistical Research Center, SPIN-UP developed a survey sent to all 759 departments in the United States that grant bachelor's degrees in physics. The survey yielded a 74% response rate distributed broadly across the spectrum of U.S. physics departments.

The site visit reports provided specific insight into what makes an undergraduate physics program thrive. In very compact form, these departments all have

- A widespread attitude among the faculty that the department has the primary responsibility for maintaining or improving the undergraduate program. That is, rather than complain about the lack of students, money, space, and administrative support, the department initiated reform efforts in areas that it identified as most in need of change.

- A challenging, but supportive and encouraging undergraduate program that includes a well-developed curriculum, advising and mentoring, an undergraduate research participation program, and many opportunities for informal student-faculty interactions, enhanced by a strong sense of community among the students and faculty.

- Strong and sustained leadership within the department and a clear sense of the mission of its undergraduate program.

- A strong disposition toward continuous evaluation of and experimentation with the undergraduate program.

In Chapter 4 of this report, each of these themes is further analyzed and illustrated with examples from the site visit departments. Chapter 5 provides a summary of the survey results.

Financial support for project SPIN-UP was provided by the ExxonMobil Foundation, the American Association of Physics Teachers, the American Physical Society, and the American Institute of Physics.

Chapter 1:
Introduction

Undergraduate physics is the miner's canary for all undergraduate science, technology, engineering, and mathematics (STEM) programs. The number of bachelor's degrees awarded in physics in the United States began a steady decline early in the 1990s. The other STEM disciplines (with the notable exceptions of psychology and the life sciences) experienced similar declines later in the decade. The reasons behind these declines are complex. The list might include the end of the Cold War and the concomitant decline in federal defense spending, changing expectations and attitudes of students, the rise of the "dot-com" enterprises, changes in secondary-school preparation of students going on to college, and a mismatch between science faculty and student expectations. For physics, recognizing the emergence of new sub-areas, such as computational physics, biophysics, and materials physics, indicates that there is a disconnect between the standard undergraduate curriculum and how physics is currently practiced. Not only are the reasons complex, they are ultimately unverifiable. This report focuses on another issue: Amidst the general decline in the number of undergraduate physics majors, a significant number of physics departments either increased substantially the number of majors in their undergraduate programs or maintained a number of majors that kept them in the top 10% or so of departments with large numbers of majors.

What makes these "thriving" departments different from those departments that experienced substantial declines? Do they have curricula that are substantially different either in content or pedagogy from those departments that have lost majors? Do their institutions make special efforts to recruit physics majors from high schools? Do the institutions draw from a body of student applicants that happens to contain more potential science majors? Do they have special laboratory and research facilities that attract physics majors? Do they make extensive use of information technology that may be attractive to potential majors? The answer to all of these questions turns out to be—by and large—"no." What then are these thriving departments doing differently? The answer to that question is what this report is about. The evidence is drawn from site visits to 21 undergraduate physics programs that, according to criteria specified by the Task Force and described in this report, have "thriving" programs and from a survey sent to all 759 colleges and universities in the United States that offer bachelor's degrees in physics.

Caveats

Before we launch into a discussion of the survey and the site visits, several caveats are in order. First, we did not attempt to measure the physics knowledge of the students in the site visit departments. A skeptic might argue that these departments have attracted more majors by "watering down" the curriculum or by "lowering standards." We saw no evidence of this in our site visits either in the courses being taught or in the statistics provided by the departments indicating that their majors follow the general patterns of graduate school enrollment and employment seen across the country. Second, we make no claims that our site visit departments exhaust the list of "thriving" undergraduate programs in the country. In fact, we had plans to visit several additional departments but could not work out mutually agreeable schedules during the 2001–2002 academic year. Along the way, we learned of several more departments that have recently revitalized their undergraduate programs and that have evidence of success. We do believe, however, that we visited a sufficiently wide range of institutions to have evidence that what we have learned has general validity.

The third caveat is that we were, because of scheduling difficulties, unable to include thriving Historically Black Colleges and Universities (HBCUs) among the departments we visited. Several HBCUs are well known for doing an excellent job of attracting physics majors and satisfy most, if not all, of our criteria for a "thriving" undergraduate physics program. However, difficulties in arriving at mutually satisfactory schedules prevented us from adding those institutions to our site visit list beyond a "tag along" visit to Xavier University in New Orleans as part of the PhysTEC program site visits. We return to the issue of diversity in physics in Chapter 6.

Undergraduate Physics in the United States

The landscape of undergraduate physics in the United States is in some ways highly heterogeneous and in other ways relatively homogeneous. Certainly the sizes and shapes of physics departments show a broad distribution. Among the 1376 four-year colleges and universities in the United States, 759 offer bachelor's degrees in physics. Many of these have very small physics programs with only one or two faculty members. Many are of modest size with four to eight faculty members. One-hundred and seventy-three institutions offer the Ph.D. in physics. Among these institutions are some of the strongest physics research departments in the world. Some of the largest physics departments have 70 to 80 faculty members. Some physics departments include astronomy and astrophysics. In other institutions, these are separate enterprises. In some colleges, physics is part of a combined physics-chemistry department, or part of a Department of Natural Sciences. The most up-to-date statistics on physics departments are available through the American Institute of Physics Statistical Research Center (*www.aip.org*).

The commonality among physics departments lies in the physics curriculum. Most college-level introductory physics courses across the country cover a common set of standard topics, usually in a one-year course (two semesters or three quarters), including classical mechanics (roughly the first half of the course), and electricity and magnetism (roughly the second half). These courses are generally taught in the traditional lecture/lab/recitation format. A mix of "modern physics" topics, including special relativity and quantum physics, is often covered in an additional semester or quarter. The "core" upper-level courses (advanced mechanics, advanced electricity and magnetism, and quantum mechanics) are even more homogeneous with a relatively small number of standard textbooks used across the country. This homogeneity in curriculum is somewhat surprising because, unlike chemistry and engineering, the physics community has no formal certification or accrediting program for undergraduate programs. The situation in physics is more akin to that in mathematics in which the community of faculty has over the years reached an informal consensus about what constitutes the core of an undergraduate program. The undergraduate physics program, at least for those students who are considering graduate work in physics, is remarkably uniform.

To complete the portrait of undergraduate physics in the United States, we need to note some further statistics. About 50% of undergraduate physics majors go on to graduate school, about 30% in physics and 20% in other fields. At the introductory physics level, annually about 350,000 students take introductory physics across the country. This number has tracked the general college enrollment for many years. About half of these students take calculus-based physics. Among those in the calculus-based physics course from which most physics majors are recruited, only 3% take another physics course. So, by and large, introductory physics is a service course at most colleges and universities.

About 20 to 30% of students who take college-level introductory physics in the United States do so in 1,600 two-year colleges. The two-year college system provides the science education for many pre-service teachers and many minority students as well. Although this report focuses on undergraduate physics programs at bachelor's degree granting institutions, we note that the contributions of two-year colleges (TYCs) to undergraduate physics education are important. Physics in TYCs is currently (2002–2003) being studied by project SPIN-UP/TYC funded by the National Science Foundation.

At the high school level, which of course plays an important role in bringing physics to the public and in preparing the next generation of physics majors, the fraction of students taking physics has been gradually increasing over the past decade, from a level of about 20% in 1990 to almost 30% in 2002. Even more noteworthy, high school physics now has a gender balance of 50:50 men and women.

The Report

The following chapters of the report describe the recent history that led to the establishment of the National Task Force on Undergraduate Physics, the procedures used in the site visits, the analysis of the site visit reports, a brief look at the results of the nationwide survey of physics departments, and an opinion piece that attempts to draw broad conclusions from SPIN-UP. Several appendices include information on physics education resources, materials used in preparation for the site visits, lists of the site visit team members, lists of presentations and articles about SPIN-UP, the report of SPIN-UP's formative evaluator, and the short site visit "case study documents," which summarize the site visit reports.

Chapter 2:
History of the National Task Force on Undergraduate Physics

1995–1998

The recent history that led to the founding of the National Task Force on Undergraduate Physics and the SPIN-UP project can be traced to 1995 when Karen Johnston, then President of the American Association of Physics Teachers (AAPT), organized a strategic planning retreat for AAPT's Executive Board. Russell Hobbie of the University of Minnesota served as facilitator for the meeting. As a result of the Executive Board's discussion, the group decided that AAPT's strategy for the next few years should focus on undergraduate physics.

The physics community had begun to notice that the decline in the number of undergraduate students receiving bachelor's degrees in physics during the early 1990s was not just a statistical fluctuation. By mid-decade, the number of physics bachelor's degrees awarded dropped to a level not seen since the 1950s. The total number of college students, by comparison, had more than doubled since the 1950s. Moreover, the decline was not evenly distributed across physics departments. Collectively, Ph.D. and masters-granting institutions suffered steeper declines than did four-year colleges. And even among those categories, there were departments that had in fact increased the number of majors or maintained an already high number. Similar declines were seen in mathematics, the other physical sciences, and most fields of engineering.

The decline in the number of majors had many causes about which we will not speculate in this report. But it is important to note that the field of physics itself has changed with major efforts in biophysics, geophysics, materials science, computational science, and physics education research, most of which have no representation in the standard undergraduate physics curriculum.

At the same time, many physics departments reported that their colleagues in engineering and the life sciences were expressing dissatisfaction with introductory physics. These departments, in most institutions, send by far the largest number of students to introductory physics. As a benchmark, we had mentioned previously that less than 3% of the students who take introductory calculus-based physics, the course most likely to be taken by potential physics majors, ever take another physics course. In other words, more than 97% of the students in introductory physics courses should be viewed as "service" students.

Of course, undergraduate physics had not been totally ignored prior to 1995. Meetings of the American Association of Physics Teachers have always had a significant number of sessions devoted to undergraduate physics. Through the efforts of Project Kaleidoscope, funded by the National Science Foundation and the ExxonMobil Foundation, a number of regional workshops on undergraduate physics had been held across the country. PKAL's Faculty 21 program, designed to promote leadership among new science faculty, included a significant number of physics faculty members. With its historical ties to the Independent Colleges Office, PKAL drew its audience mostly from undergraduate institutions or institutions with small graduate programs. The major research universities collectively were not very engaged in these efforts.

On another front, the Introductory University Physics Project, funded by NSF from 1988 through 1995 and led by John Rigden (American Institute of Physics) and Donald F. Holcomb

(Cornell University), promoted the development and testing of four new calculus-based introductory physics curricula. Sixteen colleges and universities were involved in testing and assessing the new courses [Coleman, Holcomb and Rigden, 1998]. In spite of these efforts and those of many other curriculum development projects (See Appendices I and II), most undergraduate physics programs in the 1990s closely resembled those of the 1960s.

At the national level, the entire area of undergraduate science education was under review in the 1990s. In the fall of 1996, the National Science Foundation released the results of a comprehensive review of undergraduate science, mathematics, engineering, and technology (SMET) education [George, et al., 1996]. (The previous such study was done 10 years earlier.) The primary imperative of the "Shaping the Future" report is that

> "...all students [should] have access to supportive, excellent undergraduate education in SME&T, and all students [should] learn these subjects by direct experience with the methods and processes of inquiry."

"All" in this case includes not only our physics majors, but also students in service courses, including engineers, pre-medical students, and pre-service teachers. "All" also means that we need equal access to SMET education for women, minorities, and others underrepresented in the scientific community. The "Shaping the Future" report was intended to guide both the NSF and administrators in colleges and universities across the country in examining undergraduate science education programs in the years to come. This examination presented an important challenge to the physics community: Are physics programs accessible to and effective for *all* students, and do they provide students with direct experience with the methods and processes of inquiry?

Following up on the AAPT strategic planning meeting, Robert C. Hilborn, the succeeding President of AAPT, organized a September 1996 meeting of 22 physicists and three representatives from mathematics, engineering, and chemistry to consider the current state of undergraduate physics and to recommend future directions for the physics community. The report from that meeting "Physics at the Crossroads" is available through the AAPT website (*www.aapt.org*). What emerged from the discussion was a clear vision of the need for effective action for innovation and revitalization in undergraduate physics education. Undergraduate education occupies a central position in physics: it not only has the responsibility for educating the next generation of research physicists, but must be an effective part of the science education of all students, including future K–12 teachers. Hence, undergraduate education is a major responsibility, which the physics community cannot ignore.

The group urged AAPT and APS to have the May 1997 Physics Department Chairs' Meeting focus on undergraduate physics. The steering committee for the chairs' meeting, headed by Roger Kirby of the University of Nebraska–Lincoln and Jerry Gollub of Haverford College, developed a program that highlighted the issues facing undergraduate physics. The meeting, held at the American Center for Physics in College Park, MD, drew the largest attendance of any physics department chairs' meeting. The proceedings of the conference are available through the AAPT website.

Following up on the 1997 Department Chairs' Meeting, Hilborn, Ruth H. Howes (Ball State University), and James H. Stith (then at Ohio State) decided to organize a topical conference under the auspices of AAPT and APS on "Building Undergraduate Physics Programs for the 21st Century," held in October 1998 in Arlington, VA, with support from the National Science Foundation. The meeting, modeled on Project Kaleidoscope's successful workshops on undergraduate science education, asked physics departments to send teams of two or three physics faculty members for a three-day workshop. The meeting focused on undergraduate physics as a program with many components: recruiting and retaining students, providing a stimulating and challenging curriculum, engaging students in research, building a sense of community among physics faculty and students, and so on. The meeting drew 250 participants from about 100 different physics departments. (An additional 250 applicants had to be turned away for lack of meeting space.) The report from that conference is available at the AAPT website.

1999–Present

Hilborn and Howes, recognizing the strong response to the 1997 physics department chairs' meeting and the 1998 workshop, decided that a more formal organization was needed to promote attention to undergraduate physics. After considerable discussion with physics colleagues including leaders at AIP, APS, and AAPT, they proposed the establishment of the National Task Force on Undergraduate Physics (NTFUP) as a joint effort of the three physics professional organizations. The organizations agreed to contribute $5,000 each for the initial work of the Task Force. The strategic goal for the Task Force was to "revitalize" undergraduate physics in the United States. In practical terms, "revitalization" means developing creative and constructive responses to the changing environment in which undergraduate physics operates.

The Task Force was charged with four missions:

1. To provide an overview of undergraduate physics revitalization efforts and to coordinate the efforts of physics professional organizations, individual physicists and physics departments, and funding agencies.

2. To identify areas in which revitalization efforts are needed and to catalyze projects addressing those needs. Some of the projects will be national in scope; some local, some regional. Some will be centered in universities; some in professional societies. Some will require extensive external funding; some will leverage local resources. The Task Force should encourage coordination among many groups with activities in undergraduate physics.

3. To raise the visibility of undergraduate physics revitalization by having its members speak and write about the revitalization effort and maintaining communications with the entire physics community.

4. To develop contacts with undergraduate revitalization efforts in the other scientific disciplines and to promote physics as a model for undergraduate revitalization efforts.

The Task Force was to be a relatively small volunteer group of 11 physicists from a variety of institutions: two-year colleges, four-year colleges, and research universities. The members would be formally appointed by the executive officers of AAPT and APS and the Director of Resources of AIP for two-year terms with the understanding that the appointments would be normally renewed as long as the member wishes to continue to serve. The Task Force would operate as an independent group, but would report annually to the three organizations.

The Task Force membership was recruited during the fall and winter of 1999. The first appointees were

- J. D. Garcia, Professor of Physics, University of Arizona, former program officer at NSF
- Robert C. Hilborn, Chair, Amanda and Lisa Cross Professor of Physics, Amherst College, former President of AAPT
- Ruth H. Howes, Deputy Chair, George and Frances Ball Distinguished Professor of Physics and Astronomy, Ball State University, former President of AAPT
- Karen Johnston, Professor of Physics, North Carolina State University, former President of AAPT
- Kenneth S. Krane, Professor of Physics, Oregon State University, former program officer at NSF, PI of the New Physics Faculty Workshops program
- Laurie McNeil, Professor of Physics, University of North Carolina at Chapel Hill
- Jose P. Mestre, Professor of Physics, University of Massachusetts–Amherst
- Thomas L. O'Kuma, Professor of Physics, Lee College, former President of AAPT
- Douglas D. Osheroff, Professor of Physics, Stanford University
- Carl E. Wieman, Distinguished Professor of Physics, JILA, University of Colorado
- David T. Wilkinson, Professor of Physics, Princeton University

In December 2000, Karen Johnston resigned from the Task Force to pursue other professional responsibilities. She was replaced by S. James Gates, John S. Toll Professor of Physics, University of Maryland. We note, with sadness, the untimely death of David Wilkinson in September 2002.

Ex Officio members of the Task Force are

- James H. Stith, Vice President, Physics Resources, American Institute of Physics
- Jack Hehn, Director, Education Division, American Institute of Physics
- Judy Franz, Executive Officer, American Physical Society
- Fred Stein, Director of Education and Outreach Programs, American Physical Society
- Bernard V. Khoury, Executive Officer, American Association of Physics Teachers
- Warren Hein, Associate Executive Officer, American Association of Physics Teachers
- Jeanne Narum, Director, Project Kaleidoscope

In setting up the Task Force, Hilborn and Howes articulated three important principles that underlie the Task Force efforts:

1. **Revitalization is more than curricular reform.** The Task Force efforts should differ substantially from those large-scale curriculum projects supported by NSF in mathematics, chemistry, and engineering because the Task Force will focus on the entire program of an undergraduate physics department—rather than solely on curriculum and pedagogy in introductory courses. The department's program includes recruiting and mentoring students, engaging them in research, paying attention to student learning for all students, particularly those in the other sciences, those who do not intend to be science majors, and those who intend to be K–12 teachers. The program emphasizes the department's interactions with students in class, in the research lab, in advising and mentoring, and as team members in departmental efforts such as outreach to the public and to K–12 schools.

2. **The department is the critical unit for change in undergraduate education.** Individual faculty members, of course, develop the ideas and carry out the activities, but the support of a large fraction of the department is crucial if the changes are to have lasting impact. Institutional support is also important, but the action takes place at the departmental level. Consequently, the Task Force has made a major effort to include departments of all types of undergraduate institutions ranging from two-year colleges through major research universities.

2. **All reform is ultimately local.** The Task Force recognizes that "one size does not fit all" for serious educational innovation. The Task Force hopes to identify a set of principles that are common to successful physics departments, but there is a wide diversity of approaches in applying those principles to the local situation. Each department must identify its local mission and the resources needed to carry out that mission.

In December 1999 the ExxonMobil Foundation awarded the Task Force a $25,000 planning grant to support its activities during the first year of operation. The Task Force held its first meeting in January 2000 at the AAPT winter meeting in Kissimmee, FL. The Task Force meeting focused on a broad discussion of undergraduate physics and the role the Task Force might play in addressing the challenges facing undergraduate physics.

The Task Force met for a second time in July 2000 in conjunction with the Project Kaleidoscope meeting in Keystone, CO. At that meeting, NTFUP initiated several activities. The Task Force began initial plans for a program of site visits to "thriving" undergraduate physics programs. In the fall of 2000, The Task Force conducted two pilot site visits to the physics departments at North Carolina State University and the Colorado School of Mines, both of which have thriving undergraduate physics programs. In addition, members of the Task Force accompanied the site visit teams for the APS/AIP/AAPT K–12 teacher preparation project (PhysTEC) on visits to Xavier University and Oregon State University. These pilot site visits allowed the Task Force to compile a list of the characteristics of a successful undergraduate physics program and to identify the essential elements needed for change in physics departments. A protocol for the site visit teams also was developed.

The Task Force leadership then wrote a proposal to the ExxonMobil Foundation to extend the site visits to an additional 20 or so physics departments. The project also would include a survey, carried out in collaboration with the Statistical Research Center of the American Institute of Physics, of all undergraduate physics programs in the United States. In the summer of 2001, the ExxonMobil Foundation awarded $133,000 to the Task Force's project Strategic Programs for Innovations in Undergraduate Physics (SPIN-UP) to support these activities.

In addition to the SPIN-UP activities, NTFUP agreed to serve as the steering committee for the AAPT/APS/American Astronomical Society New Physics and Astronomy Faculty Workshops, supported by the National Science Foundation. This highly successful program, targeting new tenure-line faculty, has just begun its seventh year of operation with more than 95 applications received for the planned 65 participant slots. (With renewed funding, the workshop program will be expanded to accommodate about 80 participants.) The project also will provide follow-up activities at APS divisional meetings where many new faculty members present the results of their research.

The Task Force also initiated planning for a conference on the introductory calculus-based physics course to be held during the fall of 2003. Co-chairs Bob Beichner of North Carolina State University and Ramon Lopez of the University of Texas at El Paso identified a steering committee and developed plans for a conference involving teams from university departments, probably for about 250 participants. The Task Force will be heavily involved in the conference, but leadership will be drawn broadly from the physics community. AAPT has received funding for this conference from the National Science Foundation.

We have already mentioned the project Physics Teacher Education Coalition (PhysTEC), organized through APS, AAPT, and AIP with funding from the NSF and the Fund for the Improvement of Post-Secondary Education. Three of the PIs on the PhysTEC project are ex officio members of the Task Force (Fred Stein, Warren Hein, and Jack Hehn). This program is designed to aid physics departments in working with their schools of education (or equivalent programs) to improve the science education of future K–12 teachers.

Comparison with Efforts in Mathematics, Chemistry, Engineering

The Task Force focus on the departmental undergraduate program has a flavor rather distinct from the large-scale undergraduate "reform" efforts in mathematics, chemistry, and engineering. In mathematics, the calculus reform effort, begun in the late 1980s and lasting nearly a decade, focused on new ways to teach introductory calculus. The effort was supported by more than $30M in grants from the National Science Foundation. The program led to active curricular discussions and great controversy within the mathematics community. The innovations in calculus teaching have led to the publication of several widely used textbooks, and even "mainstream" calculus texts have adopted many of the features of the reform textbooks. For details, see the Mathematical Association of America report *Assessing Calculus Reform Efforts* (1995). More recently NSF has invested about $30M in the VIGRE (vertical integration of graduate research and education) in mathematics, which links undergraduate research, graduate student support, post-doctoral support and new faculty support at about 30 major research institutions.

In chemistry the focus has been on developing curricula for the college-level introductory chemistry course through the work of five large consortia. This work was begun in 1994 and as of this writing, the work on these curricula is not yet finished. Some field-testing of the various curricular components is under way. A progress report can be found in *C&EN News*, Oct. 28, 2002, pp. 35-36 (*http://pubs.acs.org/cen*). More details about the consortia can be found at the NSF Division of Undergraduate Education website:

http://www.her.hsf.gov/ehr/due/awards.cheminit.as.

In the engineering community, NSF funded seven Engineering Coalitions aimed at attracting more undergraduate students into engineering. As of this writing, the work on these projects is still under way.

The physics community's experience with IUPP and discussions with dozens of colleagues in physics and other STEM fields led Hilborn and Howes to the conclusion that the physics community was not ready for a large-scale curricular initiative analogous to the calculus reform effort. They realized that a focus on a department's total undergraduate program—courses, undergraduate research, recruiting, retaining, advising, mentoring, physics club, etc.—was crucial for making a widespread and lasting impact on undergraduate physics. Once a department developed a strategic plan and engaged a good fraction of its faculty in carrying out that plan, the department would naturally be led to look at new textbooks and new pedagogy for its courses. Hilborn and Howes realized that new courses and pedagogy were of themselves not sufficient to "revitalize" a department's program. As the analysis in Chapter 4 of this report supports, in almost all cases, the interactions between faculty and students and among students outside normal classroom times are as important as curricular developments in the thriving undergraduate programs. A collective sense of responsibility for the undergraduate program amongst the faculty is also important.

As another part of the background for project SPIN-UP, we need to mention the important efforts in physics education research (PER). PER is the physics subdiscipline that studies how students learn (or don't learn) physics and how to translate that information into effective means for teaching physics. This effort has been under way for more than 20 years, and at present some dozen or so graduate programs offer the Ph.D. in physics with a specialization in PER. PER has led to the development of new teaching materials based on this research and to an increasingly widespread awareness in the physics community of the complex of factors that influence students' learning of physics. For a review of some of the results of PER and its influence on curricular materials and pedagogy, see Chapter 4 and Appendices I and II of this report.

Chapter 3:
Procedures for SPIN-UP Site Visits

The Task Force planned to carry out about 20 site visits to thriving physics departments during the 2001–2002 academic year. In order to facilitate a quick startup, two members of NTFUP, Ruth Howes and Ken Krane, agreed to act as project directors in exchange for release time and support for graduate and undergraduate assistants. Working closely with Bob Hilborn, the project directors hired Charles Payne of Ball State University as the external formative evaluator for the project. The team was constituted and ready to begin work in August 2001.

Site Selection

The initial step was to have the entire Task Force identify characteristics of "thriving" physics departments to be used in selecting the departments for site visits. These characteristics are

- A large number of majors (compared to the national average)

- Satisfaction of other departments within the university

- Engagement of students in the life of the department

- Undergraduate research participation

- Lively outreach efforts, recruitment programs and so on.

No single department met every criterion, but many matched several of them. In addition, the SPIN-UP leadership considered the need for diversity in type and size of institution as well as geographic distribution. Finally, letters were sent to about 100 physicists whom members of the Task Force identified as likely members of site visit teams informing them about SPIN-UP and asking if they would be willing to participate in a site visit.

The project directors developed a letter to be sent to department chairs explaining the purpose of the site visit and a letter to the site visit team, a questionnaire for the department to complete before the visit, and a protocol for the site visit. Departments were asked to pick up local expenses for the site visit and to sign a contract demonstrating their willingness to host the team. Appendix V contains the relevant letters and documents.

Site visit teams consisted of three academic physicists, who were chosen by the project directors. In so far as possible, the teams were balanced in terms of gender, ethnicity, and expertise in physics or physics education. Ideally, each team had an expert in some aspect of physics education and a faculty member active in research. In addition, each team had one member from an institution similar to the one being visited to provide perspective on administrative matters, budget, and local conditions. A member of the Task Force led each team. Generally, the team leader was identified first, and the remainder of the team was selected to balance that person's strengths. Whenever possible, teams were selected to minimize travel. This became particularly important in the immediate aftermath of Sept. 11, 2001, when air travel was difficult. The department chair and all members of the team were provided with contact information for everybody concerned.

Site Visit Information and Schedule

The site visit team and the project directors received the questionnaire report from the department at least a week before the visit. In some cases, the project directors or the leader of the site visit team contacted the department chair with additional questions. In nearly all cases, the site visit team members communicated by email or phone to discuss the upcoming visit. Each team was reminded that the site visit was not an accreditation visit, but a study of what the department was doing right.

Although individual site visit schedules varied, site visit teams usually arrived at the department in the late afternoon. The team had a dinner meeting with the department chair and/or other faculty members, particularly the director of undergraduate programs in large departments. Informal discussion at dinner allowed the department to set a tone for the visit, to discuss the schedule and to explain what the local faculty members considered important about the undergraduate program. The site visit team had the opportunity to explain SPIN-UP once again.

The next day was devoted to discussions with faculty, administrators, and students. In so far as possible, formal presentations were held to a minimum because most demographic material was already covered in the written report submitted by the department. In all cases, the team met with physics department faculty members and with physics students. Usually the team met with at least one college or university administrator. The department selected the administrator most closely associated with the undergraduate program. Frequently, the department used the site visit to publicize its undergraduate program on its own campus. We also offered any department that wanted one a colloquium by one of the team members. Large departments generally did not take advantage of this. However, it was popular among small departments, which frequently could tap funds to support the visit if it involved a public presentation. In many cases, the team met with students enrolled in service courses or with pre-service teachers. Particularly in smaller institutions, the team interviewed faculty members from other disciplines. Breakfast and lunch were generally working meals. The visit closed with a brief executive session of the site visit team followed by an exit interview with the chair, the director of the undergraduate program, or the entire physics faculty.

Following the visit, the site visit team prepared a written report for the department. Generally, the chair of the team wrote the first draft of the report, which was then re-crafted and approved by all members of the team. The report was sent to the department for correction of errors of fact and then submitted to the department chair and the Task Force. All reports are confidential. The department chair, however, could share the report at his or her discretion. The reports were generally thoughtful critiques of what made the department's undergraduate program successful. Many of them contained suggestions and comments. They followed no set format.

After receiving the written report, the project directors extracted material from the report and the department's response to the questionnaire to prepare a "case study": a formal presentation of what the department is doing and how they managed to do it, as well as steps taken to bring about change. The case study was approved for publication by the chair, who provided pictures to illustrate the online version. Twenty-one case studies appear on the AAPT website and will be included in the hard-copy version of this report. In the spirit of the site visits, the case studies highlight only the positive aspects of the department's undergraduate programs.

All members of the site visit team and the department chair were asked to fill out an open-ended evaluation of the site visit. The departments universally perceived the visit as a positive experience. Many of them stated that the most useful aspect was the time the department spent thinking about its own undergraduate program. Site visitors generally enjoyed seeing what was happening in another department and felt that the visit had been a useful experience for them. The size of the teams was considered appropriate, but in some cases, the visits seemed too short. This was particularly true in large departments. In general, the major critique of the scheduling was not having enough time to talk with students. Site visitors also emphasized the difference between SPIN-UP site visits and the usual accreditation visits.

That SPIN-UP could complete 21 site visits within one academic year represents a remarkable commitment to undergraduate education by a large segment of the physics community: the 21 site visit departments and over 70 faculty members who made up the site visit teams. SPIN-UP funds covered travel expenses for the site visit team members. The host departments paid all local housing and meal costs. Including accommodation expenses provided by the host departments, the volunteer time of the site visit teams and the time spent preparing the reports, we estimate that actual and in-kind contributions for the SPIN-UP project from the physics community are more than $130,000 beyond the funding received from the ExxonMobil Foundation.

List of Site Visits

1. Angelo State University, San Angelo, TX, *Feb. 7–8, 2002*
2. University of Arizona, Tucson, AZ, *Jan. 28–29, 2002*
3. Bethel College, St. Paul, MN, *May 2–3, 2002*
4. Brigham Young University, Provo, UT, *Nov. 15–16, 2001*
5. Bryn Mawr College, Bryn Mawr, PA, *April 15–16, 2002*
6. Cal Poly State University, San Luis Obispo, CA, *March 7–8, 2002*
7. Carleton College, Northfield, MN, *May 12–13, 2002*
8. Colorado School of Mines, Golden, CO, *Oct. 5–6, 2000*
9. SUNY Geneseo, Geneseo, NY, *April 25–26, 2002*
10. Grove City College, Grove City, PA, *Oct. 25–26, 2001*
11. Harvard University, Cambridge, MA, *Dec. 9–10, 2001*
12. University of Illinois, Urbana-Champaign, IL, *Nov. 12–13, 2001*
13. University of Wisconsin–La Crosse, WI, *March 6–7, 2002*
14. Lawrence University, Appleton, WI, *April 17–18, 2002*
15. North Carolina State University, Raleigh, NC, *Oct. 9–10, 2000*
16. North Park University, Chicago, IL, *Nov. 29–30, 2001*
17. Oregon State University, Corvallis, OR, *May 19–20, 2002*
18. Reed College, Portland, OR, *Feb. 20–21, 2002*
19. Rutgers University, Piscataway, NJ, *Dec. 3 –4, 2001*
20. University of Virginia, Charlottesville, VA, *Feb. 28–March 1, 2002*
21. Whitman College, Walla Walla, WA, *April 25–26, 2002*

Chapter 4:
Analysis

This section contains the analysis of the site visit reports. Here we extract the features that we believe distinguish a thriving undergraduate physics program from one whose performance is less than stellar. For each of the conclusions, we back up our statements with evidence from the site visit reports. The examples were chosen from the site visit reports to give some sense of the breadth of activity in the departments we visited. The examples used are not intended to endorse a particular activity as the "best practice" for a particular feature. As we mentioned previously, almost all of the site visit departments were exemplary in almost all of the features we describe. We need to emphasize, however, that it is difficult to establish a precise cause-and-effect relationship for any of the features taken individually. The collective effect, on the other hand, is striking.

General Comments

Before going into the details of the analysis, we make several important general comments:

1. There is no evidence for a single "magic bullet"—one action or activity or curricular change—that will make an undergraduate physics program thrive. In fact, it is the interaction of many activities that seems to be the key feature. Most struggling departments have some of the features identified in the thriving departments, but the interactions and the focus on undergraduate physics are lacking.

2. It has taken several years for departments that were not thriving to initiate changes and to build a thriving program. Changes take time to settle in and to make an impact.

3. Most of the crucial features do not require major external funding. The critical resource is personnel—dedicated and energetic and persevering—with a vision for a thriving undergraduate physics program. This vision is understood and clearly articulated, not only within the department, but in the institution's administration. Nevertheless, we don't wish to downplay the importance of resources: The department must have at least modest resources, both financial and human, that will allow for experimentation with the curriculum and support for student research, a physics club, and so on.

4. It is important to emphasize that none of the thriving departments have "watered down" their undergraduate programs to attract and retain majors. The site visit teams made no attempt to measure student learning directly. The teams did look at indirect evidence of what students have learned:
   (a) the quality and sophistication of student research projects,
   (b) employment of graduates, and (c) admission to graduate programs in physics or closely related fields By these indirect measures, the site visit departments seem to have rigorous curricula that prepare their students well for a variety of careers. Some of the thriving departments seem to recruit many majors from would-be engineers, mathematicians, or computer scientists just because the physics program is viewed as intellectually challenging. The key element is the sense of community that the faculty and students have established. The faculty and students work together to see that

the students benefit from the challenging curriculum.

4. Although we believe that the 21 site visit departments indeed have thriving undergraduate programs, we do not claim that these are the only such departments. Our search for thriving departments turned up at least another dozen or two departments that we would have been delighted to visit if we had had the time and resources. Furthermore, we do not claim that these site visit departments are "perfect" or "ideal" departments. Nor would the departments make such claims. They all recognize that there remains room for improvement even in the most successful programs. In addition, as we emphasize in several places in this report, what works for one institution may not be appropriate for another.

As we read through the site visit reports, we quickly realized that a relatively short list of common elements characterized the thriving departments. These elements can be expressed in several ways. First in broad categories, we recognized:

- A supportive, encouraging, and challenging environment for both faculty and students characterized by professional and personal interactions among faculty and students and among students both in class and outside class. The students expressed a strong sense of belonging to the professional physics community.

- Energetic and sustained departmental leadership focused on a vision of an excellent undergraduate physics program with continuing support from the institution's administration.

- A sense of constant experimentation with and evaluation of the under-graduate physics program to improve physics teaching, undergraduate research, student recruitment and advising and other interactions with students in line with the local needs and mission of the department and the institution.

An Analytic Outline

We also analyzed the reports with more specific categories. Here we give an outline of those categories. The remainder of this chapter expands this outline with examples from the site visit departments.

Leadership

1. Sustained leadership with a focus on undergraduate physics within the department. Most faculty members in the department placed a high value on undergraduate education.

2. A clearly articulated undergraduate mission and a vision of how that mission supports the mission of the institution. The vision is shared among the faculty and communicated to the students.

3. A large fraction of the departmental faculty actively engaged in the undergraduate program.

4. Administrative support from the dean/provost for the department's undergraduate efforts.

Supportive, Encouraging and Challenging Environment

1. Recruitment program either with high school students or with first-year students at the institution.

2. A strong academic advising program for physics majors that actively reaches out to the students.

3. Career mentoring: an active effort to make students (particularly beginning students) aware of the wide range of careers possible with a physics degree. For upper-level students the mentoring includes advice on how to apply for jobs, graduate schools, etc.

4. Flexible majors' program: Several options or tracks leading to the bachelor's degree are available (and promoted).

5. 3/2 dual-degree engineering programs, particularly at four-year colleges without engineering departments.

6. Mentoring of new faculty, particularly for teaching.

7. Active physics club or Society of Physics Students chapter.

8. Student commons room or lounge.

9. Opportunities for informal student/faculty interactions.

10. Alumni relations. The department keeps in contact with alumni, keeps data on careers of alumni, and so on.

Experimentation and Evaluation

1. Special attention paid to the introductory physics courses. The "best" teachers among the faculty are assigned to those courses.

2. Undergraduate research either during the summer or during the academic year.

3. Physics education research and external funding. Most of the faculty are aware of the findings of physics education research and pedagogical innovations based on physics education research. Some departments had one or two faculty actively engaged in physics education research. Some faculty members have received external funding for education projects.

In the following sections we will describe these categories in more detail, providing evidence for the importance of each of these activities.

The Elements of a Thriving Undergraduate Physics Program

Departmental Leadership

It should come as no surprise that departmental leadership is important. In most colleges and universities, faculty members work as fairly independent entrepreneurs, teaching their courses alone and developing their own research programs. They are evaluated and promoted based on their individual teaching and research efforts. There is no direct incentive from the institution or from the profession for working collectively on undergraduate physics. Even in four-year colleges (without graduate programs), there may be little collective responsibility for the undergraduate program. When the number of majors drops or the pre-med students complain about their experiences in an introductory physics course, it is easy to blame the students (who are obviously not as dedicated as we were when we were students, and certainly not as well-prepared), the admissions office (which always ignores students who are interested in science), or the economy, or lack of support from the administration. In thriving physics departments, however, there is a strong sense that the department collectively has the responsibility for shaping a thriving undergraduate physics program for the students that the institution brings to campus (not the students the department wishes it had). Often the chair or a group of faculty has taken the lead in helping the department maintain a focus on improving the undergraduate program. In larger research departments, it is often the chair for undergraduate studies. Furthermore, there is a tradition of keeping that focus even when the leadership changes hands.

It is important to note that in all the thriving departments, faculty members agreed that the undergraduate program was everyone's responsibility. Although almost all of the thriving programs had identifiable leaders, none of the thriving undergraduate programs was sustained by a "hero" operating in relative isolation.

> ▶ **Sustained leadership over the years:** The physics department at SUNY Geneseo was founded by Robert Sells (of Weidner and Sells textbook fame). From the beginning, the department enjoyed a focus on establishing and maintaining an energetic undergraduate physics program. The succeeding chairs have worked hard to maintain that focus and have helped Geneseo establish itself as one of the premier undergraduate programs in the SUNY system.

> ▶ **Leadership that revived a dying department:** The physics department at the University of Wisconsin–LaCrosse faced almost certain extinction in the late 1980s. The dean recommended and supported the hiring of a new chair from outside the university. The new chair, with support from the administration, increased and improved staffing and research activity, and restructured the curriculum. The new chair took the lead in convincing others in the department that they could have a thriving physics program. After two years of negotiations, efforts aimed at recruitment, undergraduate research, and 3/2 dual-degree programs were put in place. Subsequently, the number of physics majors increased dramatically.

> ▶ At the **University of Arizona,** the physics department head, with support from the higher administration, refocused the department's energies on its undergraduate program. The department now graduates about 22 physics majors and six engineering physics majors each year. About 25% of the undergraduate physics degrees are awarded to women, a figure above the national average.

We should emphasize that good leadership is not dictatorial. The leader(s) engages the entire department (or a good fraction of the department) in developing and sustaining the undergraduate program. The leadership is exercised more often by talking, persuading, cajoling, and more talking than by laying down fiats. And perseverance is primary. As we have mentioned many times, it often takes several years for the results of changes in the undergraduate program to make themselves felt. Effective leaders are patient and persevering, and they keep the department's eyes focused on the target over long periods of time.

Mission and Vision

A crucial part of departmental leadership is articulating the mission of the department, developing a vision of where the department needs to go, and keeping the department focused on that mission. It is too easy to say that the mission of the department is to "teach physics." The crucial notion is seeing how that mission is articulated for each individual department. What are the interests and needs of your students? What are the capabilities of your faculty and your institution? A small liberal arts college is not going to have either the numbers of faculty or the resources of a large research university. The small-college students are likely to have different career aspirations as well. A department in a school with a large engineering program is likely to have a mission different from that of a department that has a large pre-service teacher audience. Of course, a department's mission may change. For example, a department that in the past was mostly a service department for other science majors may decide to enhance its program for physics majors.

Each of the thriving departments we visited had a clear sense of its mission, and the departmental leadership helped articulate that mission. This articulation was particularly important for smaller departments as they recruited new faculty members. It is important that new hires understand the department's mission and that they are able and willing to support that mission.

▶ **Brigham Young University** maintains a modest graduate program in physics with about 25 graduate students. However, the department has made a strong commitment to undergraduate physics with an emphasis on undergraduate research because the university has 32,000 students of whom 30,000 are undergraduates. About 98% of BYU's students are members of The Church of Jesus Christ of Latter-Day Saints.

▶ The **Reed College** Physics Department emphasizes undergraduate research and independent work that supports Reed's overall emphasis on close faculty-student research collaborations. Four of Reed's physics majors have been recognized for their research work by the APS Apker Award (one winner, three finalists). All Reed students do a senior thesis project. External funding in the department has exceeded $2 million over the last decade.

Substantial Majority of Engaged Faculty

We all know of situations where a lone, energetic, and hard-working colleague initiated innovations in a course. Students seemed to enjoy and benefit from the change. But when the faculty member rotates out of the course or goes on sabbatical leave, the innovations are dropped. All of our site visits convinced us that having a large fraction of a department's faculty engaged in the undergraduate program is crucial to developing, and perhaps more importantly, to sustaining innovations that keep a program thriving. We emphatically point out that most of the departments displayed a broad spectrum in the level of engagement, and individual faculty members' engagement varied significantly over the years. There were periods of intense work, for example while revising large-enrollment introductory courses, with periods of less intense engagement while others carried the banner. But in all cases, the department as a whole took responsibility for the undergraduate program. Those faculty members who were less engaged nevertheless provided strong support for those who were, for the time being, carrying a somewhat heavier load. Most members of the department took part in discussions of what changes should occur and most took part in figuring out what was working and what needed repair.

Admittedly, the issue of engagement plays out differently for solely undergraduate institutions, in which perforce all faculty are engaged only in the undergraduate program, and research universities, in which—by necessity—substantial attention must be paid to the graduate programs, post-docs, and research that most likely does not involve undergraduates. Nevertheless, in solely undergraduate institutions, it is easy to find examples of physics departments in which there is little collective effort toward keeping the physics program thriving. Each faculty member may do a fine job teaching and doing research, but there may be little or no collective effort to keep the overall program alive and thriving.

How is this engagement sustained, particularly in light of pressure on the individual faculty member to spend more time on research, institution-wide committee work, professional society activities, not to mention home and family? Although the precise answer is difficult to provide, it seems that in the departments with thriving undergraduate programs, this sense of collective responsibility has been carefully cultivated over the years by the departmental leadership. New members of the faculty are mentored and guided to adopt this same philosophy. The faculty members of those departments meet often, and the undergraduate program is discussed routinely. We don't want to underestimate the difficulties faced by faculty in research universities. Their promotion and tenure decisions depend most heavily (if not exclusively) on their research productivity, despite increasing emphasis on teaching. The emphasis on research occurs at both the departmental and institutional levels and is re-enforced by the physics community, where the public recognition for research accomplishments overwhelms recognition for contributions to physics education. We are optimistic, however, that many research departments are beginning to recognize the importance of undergraduate education, if only to keep up the supply of future graduate students in physics. Many, in fact, are paying more attention to the broader role of physics in undergraduate STEM education.

This increased attention in physics shows up in the regular nationwide department chairs meetings that have a major focus on undergraduate physics. Some of these meetings, as mentioned in Chapter 2, are held by the physics professional organizations. The chairs themselves organize others, notably the "Mid-west Physics Chairs Meeting" and a meeting of chairs from departments with highly ranked graduate programs.

▶ At **Harvard,** the entire physics department meets to discuss issues of the undergraduate program. Curricular issues are hotly debated. Over the years, all of the faculty members teach in the undergraduate program. As one faculty member expressed it: "The faculty work hard to make the Harvard undergraduate physics program the best in the country."

▶ Six years ago the Department of Physics at the **University of Illinois** began a major revision of the calculus-based introductory physics sequence taken by physics majors and engineers. A team of eight faculty members worked on this revision over a period of several years (with some 10 faculty-semesters of released time to help the effort), building a solid infrastructure for a series of courses that faculty now enjoy teaching. At present, nearly 75% of the department's faculty members have taught in the revised course sequence.

▶ At the **University of Virginia,** about two-thirds of the physics department faculty are involved in teaching undergraduates at any one time. Most of the faculty see teaching as a significant part of their professional responsibility. The department has an undergraduate committee of five faculty members who make recommendations on changes to the curriculum and on other matters that affect the undergraduate program. A teaching committee reviews the teaching performance of all faculty members and plays an important role in the evaluation of faculty members.

Administrative Support

Having good administrative support would seem to be an obvious and easy matter. What administrator would not support the efforts of the faculty to improve an undergraduate program? Real-life administration, on the other hand, is heavily weighted with institutional history and institutional constraints. If a physics department has been producing only one or two bachelor's degrees per year for decades and the biologists and engineers are always complaining that introductory physics provides a very high and rough hurdle for their students, one can understand why the dean may be reluctant to provide more resources for what she thinks is a lost cause. Furthermore, the physics department is probably not the only department that needs serious attention. On the other hand, most deans are quite willing to support departments that have taken the initiative themselves, made some modest changes and have had some modest success. In all of the visited institutions with thriving undergraduate physics programs, we found strong administrative support for the physics department. In fact, in many cases the physics department was the dean's paradigm for curricular innovation, support of students, and good citizenship within the institution. It is not surprising that those deans were willing to provided additional faculty and financial resources for the department when the department made a convincing case for those resources. This support is a direct consequence of having the department's mission and vision aligned with that of the institution.

▶ The administration at **Lawrence University** provided the physics department with about $600,000 over the past 10 years to supplement external funding of about $2.5M from Research Corporation, the Keck Foundation, NSF, the Sloan Foundation and several other funding agencies. A significant fraction of this money has been used to develop "signature programs" in laser physics and computational physics, specialty programs that provide uniqueness and drawing power to the department's overall offerings.

▶ At **Grove City College,** the Dean and Provost reported that the physics department's dedication to good teaching in its service courses has been a major contributor to the "rise of physics" on the Grove City campus. Two faculty positions have been added to the physics department in the last nine years (making a total of five full-time faculty) to support the increasing number of physics majors and the increasing role of physics in teaching service courses to nonmajors.

Supportive, Encouraging and Challenging Environment and Recruitment

Almost all of the thriving physics departments had some form of active recruiting program. They had all realized that having a vibrant and exciting undergraduate physics program was necessary but not sufficient to bring students into the program. The students had to find out about the program; they had to have a sense that physics was a good undergraduate major to pursue, and that they would find the program accessible but challenging. Given the lack of information among high school students about what careers are supported by a background in physics, combined with a lack of experience with physics in high school (about 30% of high school students take physics), it is not surprising that physics departments need to do some recruiting. We found a wide spectrum of recruitment activities. Some departments were quite successful working directly with high school students and high school physics teachers. Some departments visited high schools; others invited the students for a Science Day on campus. Others found programs with high school students less productive.

Many departments actively recruited in their introductory physics courses by including career information, providing contacts with upper-level physics majors, and talking personally with students who showed an aptitude for physics. Some sponsored informal "get to know the department" meetings with short talks about research in the department, particularly student research, and career paths of recent alumni, all enhanced by vast quantities of pizza. Some invited potential majors to departmental picnics or softball games. Many chairs wrote letters and sent departmental brochures to all admitted students who indicated some interest in physics or whose academic records indicated that they might be potential physics majors.

Several of the site visit departments offer a one-credit-hour course ("Introduction to Physics as a Profession," for example) for first-year students aimed specifically at introducing the students to the department and to the potential careers one can pursue as a physics major. These short courses were often cited by students as being very influential in their decisions to become physics majors.

▶ The **Lawrence University** physics department invites roughly 30 "select" high school students to visit campus for a weekend workshop in February or March. Each of the students is hosted by a physics major from Lawrence and spends time doing laboratory work using research equipment at Lawrence. Approximately 30% of the workshop attendees matriculate at Lawrence. The annual cost of $15,000 is underwritten by the Office of Admissions, which handles the mailings and invitations. This recruiting effort has had a profound effect on the development of physics at Lawrence.

▶ At **North Park University**, the chair of the Physics Department has the Admissions Office send names of all prospective students interested in physics, engineering, or science to the department. The chair phones or emails all of these students and invites each prospective student and their parents personally to visit the department and follows up the visits with a personal and often humorous postcard.

▶ At **Bryn Mawr College**, students involved in the introductory physics courses are given tours of the research laboratories. Upper-level students involved in the research laboratories give presentations for these students at a mini-symposium. Many students cited the research opportunities as playing an important role in their decisions to become physics majors.

Advising

Once students declared themselves as physics majors, the thriving departments provided active advising. The advising took many forms: In some departments, one faculty member served as undergraduate advisor for all the majors, providing common information and advice, resolving scheduling problems, and checking on required courses, for example. In other departments, the advisees were spread among all the faculty. Some departments used a mixed mode with one faculty member serving as chief advisor but with all students assigned to other faculty members for additional advice. No one scheme seemed to work significantly better than the others.

In additional to formal advising, the students in the thriving programs reported to the site visit teams that faculty were available almost 24/7 for informal advising, help with homework (even for courses they were not teaching), career information, and just general talking about life. We got the sense that many of these informal discussions often dealt with course selection, how to get a summer research position, and other topics that might normally be relegated to formal advising appointments.

▶ The Undergraduate Program Director in the Department of Physics and Astronomy at **Rutgers University** handles all of the advising for undergraduate majors. The faculty and the departmental leaders believe that centralizing the undergraduate advising was the most important factor leading to the growth in the number of physics graduates (doubling from about 20 in 1980 to 40 in 2000). The students support this conclusion, expressing strong appreciation for the director's individual concern for them and for the consistency of the advice they received.

▶ At **North Carolina State University,** students declare their majors when entering the institution. The physics majors enter a special section of the introductory course with special laboratories and a unique curriculum. A small group of advisors works closely with the physics majors and follows them from freshman year forward.

Career Mentoring

Today's students have a strong interest in shaping their careers relatively early in their undergraduate years. One might argue that students have always had strong career interests, but today's students seem to be particularly vocal and focused on careers. If the students are not, certainly their parents are. Physics finds itself in an unusual situation in the sciences: Most students (and their parents) think that the only careers available to physicists are those in academe or in basic research in the national labs. In fact, less than 20% of people with a degree in physics (bachelor's, master's, or Ph.D.) pursue careers in academe or the national labs: About 30% of physics bachelors go on to graduate school in physics. Of those, less than 40% end up with Ph.D. jobs in academe or national labs. (See the AIP Statistical Research Center website for further details.) The vast majority do something else. To complicate matters, most high school physics teachers and physics faculty members in colleges and universities are only dimly aware of these (obviously misnamed) "alternative" career paths. For better or worse, most of these other jobs do not have "physicist" in the job title. Almost all of the site visit departments provide extensive career information and career counseling to their majors and potential majors. One of the most effective career advising tools is pointing to the department's own alumni. Many departments have their alumni return to give talks about their careers in industry and business as well as those who pursue academic and basic research careers.

As an aside, we note that the physics professional organizations AIP, APS, and AAPT now have available extensive information about careers pursued by people with physics degrees. Students can be directed to these organizations' websites for abundant and up-to-date career information. APS's Committee on Career and Professional Development runs a CPD liaison program in which a faculty member in a department is designated as the primary point person for APS career information.

> ▶ At **Carleton College,** prospective physics majors take a one-credit-hour course "What Physicists Do" that brings to campus alumni as well as other speakers to show how a major in physics leads to a wide range of careers.

> ▶ The **University of Arizona** physics department hosts an Academic Support Office for undergraduates. Among other functions, the office maintains an employment database where students can find information on internships as well as permanent employment. The department also maintains a webpage listing of alumni and their present activities, and a program under which alumni are invited back to give talks to the department.

> ▶ **Bethel College** maintains close ties with high-tech industries in the Minneapolis/St. Paul area and places many students in internships with these industries. (These connections often lead to equipment donations and funded research contracts, as well.) The entire physics faculty at Bethel meets to match students with available internships.

Introductory Physics Courses

For most physics departments, the large introductory physics courses are a key component in their undergraduate programs. This is where the department has its first contact with potential majors and where it provides its largest service to the rest of the institution. The economics of higher education often dictates that these courses have large sections and only a few faculty members (and often just one) assigned to teach them. All of the site visit departments work very hard at making the introductory courses as good as possible. Most assign only their "best" and experienced faculty to those courses. When new faculty members rotate into those courses, they often do so first as "apprentices" with more experienced faculty. Many faculty teaching in those courses are using innovative pedagogy such as peer instruction [Mazur, 1997], just-in-time teaching [Novak, et al., 1999], and active demonstrations [Sokoloff and Thornton, 1997] [Thornton and Sokoloff, 1998]. Few of the departments, however, would claim that they are doing anything radically different with their introductory courses. Some departments have developed special courses or special sections of the introductory course to appeal to potential physics majors.

The common feature among the site visit departments was a sense of constant monitoring and refinement of the introductory courses, both those for majors and those for nonmajors. By and large, most of the departments had a sense of collective "ownership" of the introductory courses. Although individual faculty members would tinker and adjust the introductory course when they were teaching it, no major changes were introduced without significant discussion and buy-in from the rest of the department.

▶ The Physics Department at the **University of Illinois** undertook a multi-year, massive restructuring of its introductory physics courses, which serve a very large number of engineering majors. The goal was to develop a solid infrastructure so that teaching the courses did not require superhuman efforts. Students attend lectures twice a week, submit homework on the computer, and then attend a two-hour discussion section covering the same material. The labs were reorganized to emphasize conceptual understanding based on the "predict, observe, explain" model of Thornton and Sokoloff [Thornton and Sokoloff, 1998]. Lectures are based on PowerPoint presentations so all lecturers cover the same material. T.A. training has been enhanced to prepare the T.A.s for the new type of discussion sections. In 2001, 75% of the T.A.s were rated as excellent, up from 20% in 1997. The department also added two new positions. One is a staff position to assist with the introductory courses. The other is a new administrative position—"Associate Head for Undergraduate Programs."

▶ At **Brigham Young University**, the physics department supports all the introductory courses with tutorial labs, peer student assistants, and faculty assistance with special rooms available and staffed for the introductory physical science courses and the introductory physics courses. The department maintains faculty committees to oversee the service courses and interact with appropriate departments on campus for which these courses provide support.

▶ **Carleton College** offers an unusual structure for its introductory physics course. Its one-term (10-week)-duration course is split into two half-term courses. Starting in the Winter Term, students usually take a one-half-term course in Newtonian Mechanics or, for students with sufficient high school preparation, a half-term course on "Gravitation and the Cosmos." Both sections are followed by a half-term on Relativity and Particles. The notion is to expose the students to exciting, up-to-date topics early in their careers. Other traditional introductory topics are subsumed into an intermediate-level sophomore sequence of atomic and nuclear physics, two half-term courses in classical mechanics and computational mechanics, and electricity and magnetism.

▶ At the **University of Virginia** about one-half of all undergraduates students have taken at least one course in the physics department. Many non-science majors take one or two semesters of "How Things Work" or "Galileo and Einstein" or a conceptual physics survey course. The physics department has an excellent reputation among non-science students at Virginia.

Flexible Majors' Program

Most of the site visit departments have developed a set of requirements for the major with considerable flexibility to meet the needs of students with a broad spectrum of career interests. Many programs have a set of core requirements that all majors satisfy, but they leave considerable flexibility for options at the upper level. This flexibility seems to be appearing in many physics departments across the country. Many site visit departments had explicit "tracks" for students who want to combine physics and engineering, physics and chemistry, physics and computer science, physics and biology, even physics and business. Others allow for a concentration within physics, for example lasers and optics or materials science. This flexibility is often important to students who may want additional specialization beyond the usual array of undergraduate physics courses to enhance their career options or to follow up on some scientific or technical interest beyond physics. This flexibility also reflects the current practice of physics, where some of the most exciting developments are occurring at the interfaces between physics and other scientific disciplines.

These departments have dealt with the unavoidable criticism of "diluting the major" or "making the major less rigorous" by recognizing that students who intend to go to graduate school in physics, for example, need to have taken a set of courses somewhat different from those taken by a student who intends to go to medical school. As another example, a student who intends to be a high school physics teacher is probably better served by taking some biology and chemistry courses rather than a second advanced course in quantum mechanics. The advising program plays a critical role in guiding the students in choosing the set of courses that best meets their needs.

It is important that the department treat students who don't intend to go to graduate school in physics as full citizens of the department. It is too easy to fall into the trap of saying that only people with Ph.D.s in physics are the ones who may be called "physicists." The site visit departments seemed universally to go out of their way to celebrate the diverse career paths of their students.

▶ **Harvard University's** physics department, which graduates 50 to 60 majors each year, supports two levels of majors: The basic program requires a total of 12 courses in physics and mathematics. The "honors" program requires in addition two advanced mathematics courses, an advanced lab course, and three additional physics courses. There are also several joint-major programs: physics and chemistry, physics-mathematics, physics-astronomy, physics-history of science, a biophysics option, and a physics teaching program for those intending to teach physics at the secondary school level.

▶ **Whitman College,** which graduates about 10 majors each year, has several "combined majors" programs in mathematics-physics, astronomy-physics, and geology-physics.

▶ **Oregon State** radically revised its upper-level curriculum to allow more flexibility for its many transfer students and to provide a more integrated experience for its majors. The junior year consists of nine 3-week "paradigms" on such topics as Oscillations, Vector Fields, Energy and Entropy, Waves in One Dimension, and so on. In the senior year the students take a series of more traditional capstone courses in classical mechanics, quantum mechanics, electricity and magnetism, statistical mechanics, optics, and mathematical methods. The development of the Paradigms model was supported by grants from the National Science Foundation.

▶ The physics department at **Rutgers University** offers four different options for undergraduate physics majors. The Professional Option is aimed at students who intend to go to graduate school in physics. The Applied Option and the Dual-Degree option attract students looking for more applied work in physics or engineering. The General Option is intended for students who plan careers in law, medicine, or secondary-school teaching. A new astrophysics major has recently been introduced. The department is considering adding an engineering physics degree.

3/2 Dual-Degree Engineering Programs

Many colleges without their own engineering schools are participants in 3/2 dual-degree engineering programs in which a student spends three years at the college and two years at the cooperating engineering school. The student then graduates with a B.A. from the college and a B.S. from the engineering school. In many cases, these students are physics majors. Physics departments have found that a 3/2 engineering program is quite attractive to high school students who are interested in engineering careers but who want a liberal arts background before committing themselves to a more technical career. The students may also want to have a few years to think about which flavor of engineering they want to pursue. No matter what the specific motivation, many colleges and universities without engineering programs find that a 3/2 program attracts students who would not otherwise consider their programs. Once the students are enrolled, a significant number decide to stay four years at the college and be "regular" physics majors, partly because they want to graduate with their friends and particularly because they find the physics department hospitable. Many of these students then go to graduate school in engineering or applied physics.

▶ **SUNY Geneseo** admits about 40 students each year interested in the 3/2 dual-degree engineering program. Many of these students are subsequently recruited to be physics majors, and many of them decide to finish a physics major program at Geneseo in four years and to pursue graduate studies in engineering.

▶ **Bethel College** offers both 3/2 and 4:2 (B.S. in physics, M.S. in engineering) programs and has recently instituted a major in Applied Physics.

▶ The **University of Wisconsin-LaCrosse** recently established 3/2 arrangements with the University of Wisconsin campuses in Madison, Milwaukee, and Platteville and with the University of Minnesota. About half of the graduating majors each year are in the 3/2 program.

Undergraduate Research

It is safe to say that the past 20 years have seen a revolution in undergraduate research participation. Fairly rare several decades ago, undergraduate research is found nowadays in almost all colleges and universities. These institutions and their students have recognized that participating in research where the answers cannot be found in the back of the book and where even the procedures are not initially well-defined is a powerful educational tool. It gives students a sense of what actual scientific research is like and it motivates students because they see their classroom learning in action. In addition, having students engaged in the research helps move along the faculty members' research programs, particularly at colleges without graduate programs. Most undergraduate research programs provide opportunities for the students to give public presentations of their research results. These presentations are excellent opportunities to develop the students' communication skills, important for almost all careers, and makes the students feel that they are indeed part of the scientific research community.

The 1998 Boyer report (*http://naples.cc.sunysb.edu/Pres/boyer.nsf/*) called upon research universities to achieve a greater integration of research with undergraduate education and made specific suggestions for curricular reform to achieve that end. A 2002 follow-up report (*http://www.sunysb.edu/pres/0210066-Boyer%20Report%20Final.pdf*) indicated the considerable progress that has been made in achieving the goals outlined in the earlier report. Both of these reports are available through the SUNY–Stony Brook Reinvention Center (*http://www.sunysb.edu/Reinventioncenter/*). Although these reports dealt only with research universities, they contain important lessons for undergraduate programs at all types of institutions.

All of the site visit departments had thriving undergraduate research programs. About half of them *require* participation in undergraduate research for the major. In addition to on-campus research with their own faculty, many students take advantage of off-campus opportunities, for example, in the Research Experiences for Undergraduate programs sponsored by the National Science Foundation and some of the national laboratories. In many departments, students are encouraged to participate in research even after their first and second years, just to see what research is like and to experience working on a research team. Most undergraduate research programs focus on work in the summer after the junior year and during the senior year, often culminating in a significant research thesis or report.

Undergraduate research participation benefits both the students and the department in many ways that go beyond just the completion of the research. Students gain experience working in teams and communicating their results, both orally and in written reports. The shared research experience gives the students a deserved sense of being part of the scientific community, not just passive consumers of science through their courses. Most departments recognize the importance of undergraduate research in building a sense of community within the department. In addition, the time students spend working directly with faculty members on research provides many opportunities for informal advising.

▶ **Angelo State University** physics majors are required to complete a three-hour research course prior to the fall semester of the senior year and to participate in a student research project either during the academic year or during the summer.

▶ At **Brigham Young University** two-thirds of the 28 physics faculty members are engaged with undergraduate students doing research. (The department also has a Ph.D. program with about 25 graduate students.) One faculty member serves as undergraduate research coordinator. A senior thesis, honors thesis, or capstone project is required for the Bachelor of Science degree in physics. With 45 to 49 graduates per year, the research supervision load of the faculty is fairly high. The university provides about $20k per semester to support the research of 20 to 25 students. The department also hosts an NSF-funded Research Experience for Undergraduates program during the summer. More than half of the department's B.S. in Physics and Physics and Astronomy majors gave talks at regional or national meetings last year.

▶ **Carleton College** physics majors complete a senior thesis project, which may be in an area associated with faculty research. Other thesis topics evolve out of a recently improved junior-year laboratory course (entangled photon detection and atom trapping, for example). Others focus on contemporary research topics such as LIGO or CP violation.

Physics Clubs and Commons Rooms

Almost all of the site visit departments have an active physics club or Society of Physics Students chapter. The activities of these clubs varied from college to college but they included organizing informal gatherings of students and faculty, running outreach programs to the local schools, organizing tutors for introductory physics students, inviting and hosting speakers for the physics colloquium series, talking with first-year students about becoming physics majors, providing feedback to the department about the undergraduate program, and so on. Most of the clubs have a faculty advisor, whose role is often limited to seeing that the club's activities get started each year with the students, in practice, doing almost all of the work. The benefits of having an active physics club include giving a structure for building a sense of community and responsibility among the students, inviting new students into that community, and providing many opportunities for informal interactions among the faculty and students. The students in those departments with SPS chapters enjoyed the contact with the American Institute of Physics and the regional "zone" meetings of SPS chapters from neighboring institutions. AIP provides a newsletter and career information to students in SPS chapters.

Almost all of the site visit departments provide some commons space for their majors. Sometimes the space is just the back of a classroom or a lab room that was vacant in the evenings. In most cases, the students have access to a dedicated room equipped with a computer or two, some physics reference books, and, of course, a coffee pot and microwave oven. Providing the student space signals to the students that the department takes them seriously and that they are indeed part of the department. The study sessions and physics club meetings held in that space contribute to the sense of community among the students.

▶ The SPS chapter at the **University of Arizona** is involved in a number of aspects of the Department of Physics programs, such as interviewing prospective faculty candidates, participating in outreach activities, and assisting with student orientation. The undergraduate majors have a dedicated lounge area, and an undergraduate council provides advice to the department chair and serves as a liaison between the chair and the undergraduate majors.

▶ At **Cal Poly San Luis Obispo,** the active SPS chapter helped set up a centrally located physics majors' lounge area called "h-bar." This space provides an area where informal faculty-student interactions and student-student interactions can occur. Students tutoring other students also use this room. The area has ample whiteboard space and is adjacent to the project rooms where seniors have workspace for their research activities. Students—from first-year students to seniors—attested to how they make use of this space for study groups, how the more senior students help the less experienced ones, and how the room led to remarkably high community spirit.

Mentoring for New Faculty

Most college and university faculty members start their teaching careers with little or no training in teaching. They may have served as teaching assistants while in graduate school, but particularly in the sciences, may have had no "full responsibility" teaching. As they take up their first full-time academic positions, they are hit with a wide range of unexpected responsibilities: managing grading and record-keeping for a large class, dealing with student complaints, training their own teaching assistants as well as organizing a syllabus, preparing lectures and labs and writing and grading exams. At the same time, they are working hard to get their research programs up and running. It comes as no surprise that most new faculty find the first years of teaching some of the most stressful and demanding of their academic careers. All of the thriving departments we visited had some means for working with new faculty to help them through this difficult period. Some departments had formal mentoring programs, pairing the new faculty member with a more experienced faculty member. Some sent their new faculty members to the AAPT-APS-AAS-NSF New Physics and Astronomy Faculty Workshops, held each fall at the American Center for Physics. In some departments, the chair played the role of mentor. Some colleges and universities had Teaching and Learning Centers, which provided advice and feedback for faculty. None of the thriving departments simply threw new faculty members into the turbulent waters of teaching and expected them to learn to swim on their own. In most of the departments new faculty were invited to talk about their teaching with more experienced faculty and felt comfortable doing so: not only about a good way to teach projectile motion, but how to deal with a depressed student who has stopped coming to class or what to do with an overly enthusiastic male student who tends to dominate his lab group. This sense of collaboration on teaching occurred with the full knowledge that faculty colleagues will need to make recommendations for reappointment, promotion, and tenure based on the new faculty member's teaching record.

> ▶ The head of the physics department at the **Colorado School of Mines** sends each of the new faculty members to the New Physics and Astronomy Workshops. The head has lunch with junior faculty regularly. When the new faculty members are assigned to teach the introductory courses, they first serve as "apprentices" with more senior faculty. The department has a "PET" (Peer Enhancement of Teaching) program in which new teachers trade classroom visits with experienced colleagues.

> ▶ At **Cal Poly San Luis Obispo,** the new physics faculty members are introduced to a clear set of metrics (the "Bailey list") based on a principle of "occasional external validation" against which their performance is to be measured. The presence of these clear guidelines helps provide a comfortable and "transparent" environment in which all faculty members feel free to focus upon the issues of quality instruction.

Informal Student/Faculty Interactions

We have mentioned already several ways in which informal student/faculty interactions occur. In addition to the usual array of departmental picnics and pizza parties, our site visits taught us about informal hallway conversations, informal talks by faculty about the department's research program, and an open-door policy for faculty, who encourage students to drop by for questions at any time. It is through these informal interactions that the faculty get to know their students more personally and the students get to know the faculty as people who have lives and interests outside the classroom. These personal interactions allow the faculty to give the students better academic and career advice. They also make the students more comfortable in approaching the faculty members with questions about physics, careers, and about life.

> ▶ The **SUNY Geneseo** physics majors participate in an annual bridge-building contest and a Physics Bowl attended by all the physics faculty members. The department maintains an "open door" policy and faculty members are available to talk to students about physics (even in courses they are not teaching), careers, personal issues and so on, at almost any time. Picnics, the Sigma Pi Sigma induction banquet, a junior-senior dinner in the spring, and a commencement luncheon provide opportunities for physics majors to interact with faculty and their families.

> ▶ At **Harvard,** the physics faculty have lunches at the Harvard Faculty Club for their majors as well as fall and spring departmental picnics. A former chair hosts a weekly physics study night in one of the student houses (dormitories). The undergraduate majors join the graduate students for an annual "puppet show," put on to "roast" the physics faculty.

Alumni Relations

All of the site visit departments keep in touch with their alumni. This contact serves several purposes:

- The alumni provide important feedback to the department about the strength and weaknesses of its undergraduate program as the students pursue a wide range of careers.

- The alumni serve as vivid examples of what careers can be pursued with a physics major. These examples are often important in convincing beginning students and their parents that majoring in physics will provide the students with a good background for many interesting careers.

- The alumni are often good sources of contacts for opportunities for research, internships, and jobs for the department's students.

- The alumni are often good speakers for departmental colloquia, particularly for areas outside of basic research.

- By tracking alumni career trajectories, the department has a much more realistic sense of its students" interests and how a physics major can help them to pursue those interests.

▶ In the lobby near the **SUNY Geneseo** physics department office hangs a map of the United States overlaid with photos of recent physics alumni and brief captions indicating where they are employed or the graduate school they are attending. Students at Geneseo cited this map as giving them good information about the wide range of careers possible with a physics major.

▶ At the **Colorado School of Mines,** the Department of Physics maintains an active "Visiting Committee" composed of representatives from local industry, research university faculty, and recent CSM alumni. The department also keeps in contact with alumni through a survey inspired by the Accrediting Board for Engineering and Technology (ABET) and an annual newsletter.

▶ The SPS chapter at **Bethel College** sponsors two or three talks each year by alumni working in local industry. Many local alumni also attend the annual SPS banquet. This network of alumni provides many opportunities for student internships during the summer and part-time work during the academic year.

Physics Education Research

Physics education research (PER) is a growing branch of physics research that focuses on studies of student learning and problem solving, as well as on applying findings from learning research to the development of curricular materials. Physics is considerably ahead of all the other sciences in having substantial literature on student learning and problem solving. The recent report by the National Research Council, *How People Learn: Brain, Mind, Experience and School* (1999), and a recent review of PER by McDermott & Redish (1999), highlight many of the salient findings in PER. For example, research on learning strongly suggests that active engagement on the part of students is more conducive for knowledge acquisition, recall, and conceptual understanding than more passive approaches [Bransford, Brown and Cocking, 1999]. This research has in turn led to the development of pedagogical techniques that get students more actively involved in learning physics. See, for example [Mazur, 1997], [Mestre, *et al.*, 1997], [Sokoloff and Thornton, 1997], and [Thornton and Sokoloff, 1998].

Yet, despite the progress that PER studies have brought in understanding teaching and learning, there appears to be some controversy in the physics community about the implications of PER. The controversy arises because of some of the dogmatic interpretations (really, from our point of view, misinterpretations) of the results of PER. For example, the well-documented role of interactive-engagement techniques in enhancing students' conceptual understanding of physics could be misinterpreted as saying that there is no role for the traditional lecture in physics. The true lesson, we believe, is the following: Know your students. For some students, lecturing is just fine. They will do the interactive-engagement work on their own in small study groups, for example. For other students, a mix of lecturing (which is what they expect to find in science courses) and interactive-engagement methods is best. For yet other groups, hands-on interactive-engagement without much lecturing might be the preferred mode. In all cases, knowing your students and getting feedback about what works with those students are the key features.

All of the site visit departments had some (but by no means all) faculty members who were aware of the findings of physics education research. This awareness came about through reading articles in *The American Journal of Physics* and *The Physics Teacher*, by attending meetings of the American Association of Physics Teachers, by participating in the NSF-funded New Physics and Astronomy Faculty Workshops, or by inviting faculty actively engaged in PER to give talks to the department. Most of the site visit departments were experimenting with modes of pedagogy suggested by PER as effective in enhancing student understanding of physics, but none had completely forsaken traditional lectures. Nor had any adopted wholesale the curricula that have been developed and tested by PER faculty. We did find, however, a sense of continuous experimentation and evaluation of the physics teaching, particularly in the introductory courses. The students reported enthusiastically about the energy, care, and concern expressed by the faculty in their teaching, while at the same time recognizing that the physics courses were often the most demanding courses on campus. The students sensed that the faculty members were there to work with them and to help them master the skills and develop the understanding necessary to pursue work in physics.

PER References

L.C. McDermott and E.F. Redish, "Resource letter: PER1: Physics education research," *Am. J. Phys* **67**, 755–767 (1999).

E Mazur, *Peer Instruction: A User's Manual* (Prentice Hall Upper Saddle River, NJ, 1997).

J.P. Mestre, W.J. Gerace, R.J. Dufresne, and W.J. Leonard, "Promoting active learning in large classes using a classroom communication system," in E.F. Redish and J.S. Rigden, eds, *The Changing Role of Physics Departments in Modern Universities: Proceedings of the International Conference on Undergraduate Physics Education* (American Institute of Physics, Woodbury NY, 1997), pp. 1019–1036.

John D. Bransford, Ann L. Brown and Rodney R. Cocking, eds, National Research Council, *How People Learn: Brain, Mind, Experience, and School* (National Academy Press, Washington, D.C., 1999).

D.R. Sokoloff and R.K Thornton, "Using interactive lecture demonstrations to create an active learning environment," *Phys. Teach.* **35** (6), 340–347 (1997).

R.K.Thornton and D.R Sokoloff, "Assessing student learning of Newton's Laws: The force and motion conceptual evaluation and the evaluation of active learning laboratory and lecture curricula," *Am. J. Phys.* **66**, 338–352 (1998).

▶ **North Carolina State University** has long been active in physics education research. In recent years Project Scale-Up, funded by NSF, has focused on developing means for using interactive-engagement techniques in large introductory courses. At present three sections of the calculus-based introductory physics course use the Scale-Up format. The department has recently hired two additional faculty in physics education research.

▶ As a result of its experience with revising its introductory physics courses, the **University of Illinois** Department of Physics is establishing a physics education group. The initial group includes two current faculty members and three graduate students.

▶ **Rutgers University** recently set up a physics education research group with the hiring of a senior faculty member from another major research university.

▶ The head of the physics department at the **Colorado School of Mines** actively promotes and rewards the use of innovative pedagogy at all levels of the curriculum, and most faculty are trying new pedagogy in their courses. Faculty members are encouraged to seek external funding to support pedagogic reforms.

Counter-Examples

The Task Force did visit two physics departments whose undergraduate programs turned out not to be as "thriving" as we had anticipated based on our preliminary information. We decided not to include them in the series of case studies. They do, however, form a small, but useful set of "control group" departments. These departments were by no means "bad" departments, but for a variety of reasons their undergraduate programs were not very successful. One, a large research university, graduates about 20 majors per year, but upon closer examination, we discovered that the relatively large number of majors was due primarily to the efforts of two, non-tenure-track faculty members close to retirement. Such a program is not sustainable. The other department was in a university that serves a large minority population. The department was quite successful in establishing a substantial graduate program and building up research efforts by most of the faculty. However, the focus on getting the research program established had siphoned energy away from the undergraduate program, which also suffered from lack of support in the administration. Since the site visit, several faculty members in the department have begun planning actions to revitalize (or in this case, vitalize) the undergraduate program.

Both of these counter-examples demonstrate the importance of having a clear focus on undergraduate physics and developing broad support and engagement in the undergraduate program by a substantial fraction of the department's faculty. Both these departments have the raw material for thriving undergraduate physics programs, but they lacked the focused leadership and widespread engagement by the faculty required to shape that raw material into an effective program.

Chapter 5.
The Survey

Background

Recognizing that our site visits to 23 institutions (representing 3% of the undergraduate physics programs in the United States) may not have given us a complete or representative overview of the state of U.S. undergraduate physics, we sought to augment our conclusions from the site visits with results from a survey that would provide a more complete view of the status of undergraduate physics education. To accomplish this result, the Task Force composed a survey form (see Appendix VII) to distribute to all undergraduate programs in the United States. Its goals were to provide data of greater statistical reliability than those obtained from the site visits, to give a more detailed and comprehensive snapshot of undergraduate physics programs, and to reveal trends or circumstances that might have been missed in the site visits. The distribution of the survey form and the analysis of the responses were done with the collaboration of the Statistical Research Center of the American Institute of Physics.

The survey form, which sought to gather information about curricula, recruiting, advising, alumni contacts, and reform efforts, was posted on a website. Chairs of all 759 undergraduate programs were notified on April 17, 2002, and asked to participate in the survey. They were told that the survey results would be analyzed only statistically and that we would not identify any of the respondents in our publications. Follow-up reminders were sent to nonrespondents on April 29, May 22, and June 5. Data gathering closed on June 17, at which time 561 replies had been received, representing 74% of the programs surveyed. This represents an extraordinarily high response rate, which typically indicates that the questionnaire focused on issues of concern to the respondents.

Originally there was concern that the response would not represent a fair cross-section of the undergraduate programs; we feared that small or inactive departments would be less likely to respond, as they have little to report. Tables 1 and 2 compare the response rates according to number of faculty and according to number of graduates (totaled over the three years 1998–2000). Data on department size and number of graduates in Tables 1 and 2 were collected independently by the AIP Survey Research Center as part of its regular annual survey of physics programs.

The results summarized in Tables 1 and 2 demonstrate that our fears of a skewed response rate were not justified. Table 1 shows that the survey response rate was quite constant independent of faculty size. Table 2 shows that the response rate was constant independent of number of graduates and also independent of highest degree awarded. Based on the analysis of these three factors (faculty size, number of degrees awarded, and highest degree awarded), the response rates appear to be free of bias in any particular direction. Small deviations between different categories can be ascribed to statistical fluctuations and are generally within the anticipated statistical sampling errors.

Table 1. Survey response rate according to size of department (number of faculty)

Number of faculty (FTE)	Number of departments by highest degree			Total number of departments	Response rate %
	B.S./B.A.	M.S.	Ph.D.		
≤ 2.0	95	—	—	95	67
2.1 – 3.0	87	1	—	88	73
3.1 – 4.0	87	1	—	88	78
4.1 – 6.0	108	4	2	114	71
6.1 – 9.0	80	18	8	106	73
9.1 – 15.0	34	33	27	94	83
15.1 – 25.0	5	10	57	72	74
25.1 – 39.9	4	3	46	53	81
≥ 40.0	—	—	33	33	70
Total number of departments	500	70	173	743	74

Table 2. Survey response rate according to number of bachelor's degrees awarded (1998–2000 total)

Number of bachelor's degrees (3-year total)	Number of departments by highest degree			Total number of departments	Response rate %
	B.S./B.A.	M.S.	Ph.D.		
New departments*	11	1	—	12	58
None	22	1	—	23	74
1 to 5	156	10	5	171	71
6 to 9	124	17	21	162	78
10 to 14	92	14	28	134	69
15 to 29	83	25	56	164	76
30 to 44	22	3	28	53	79
45 or more	5	—	35	40	73
Total number of departments	515	71	173	759	74
Total response rate	74	68	77	74	

*These 12 departments were added to the AIP database during the three years 1998–2000.

Courses and Curricula

The first section of the survey dealt with courses and curricula. One of the conclusions from the site visits was that thriving departments were often characterized by a diversity of physics degree programs or "tracks." We sought to use the survey to determine whether that finding, based on a limited number of data points, is broadly characteristic of the physics community. Moreover, we requested information on the alternative degrees and the number of physics credits required.

Of the responding schools, 81% characterized themselves as following the traditional semester system in awarding course credits, 6% awarded credits following a traditional quarter system, 10% awarded credits on an alternative system with one credit per course, and 3% followed various other schemes for awarding credits. We asked each school to begin by providing information about its "standard" physics program. This is usually the most rigorous program, requiring the greatest number of physics credits and often designed to prepare students for graduate study or professional work in physics. Table 3 shows the survey results for the number of physics credits required for this degree program compared with the total number required for a bachelor's degree.

Table 3. Credits required for "standard" physics degree

Academic Calendar	Total credits for bachelor's degree		Physics credits for "standard" degree		Number of respondents
	Low	High	Low*	High*	
One credit per course	30	36	8 (25%)	12 (39%)	57
Semesters	110	146	32 (25%)	50 (40%)	452
Quarters	175	196	54 (29%)	87 (48%)	30

*Low and high figures represent 10% and 90% range; that is, 10% of the respondents are below the low figure and 10% are above the high figure. The numbers in parentheses represent the fraction of the total credits represented by the required physics credits.

Institutions vary in terms of the number of total credits required for a baccalaureate degree, and they vary in terms of the proportion of those credits that must be completed in physics to be awarded a physics degree. The physics fraction of total graduation credits required for the "standard" physics degree is typically in the range of 25–39% for schools on the one credit per course system, 25–40% for schools on the traditional semester system, and 29–48% for schools on the traditional quarter system. Thus, to earn a physics degree, virtually all physics departments require that at least 25% of all credits be taken in physics and comparatively few departments require more than 40%.

We also surveyed the physics credits required for different types of "standard" degree programs. Table 4 shows these data, along with the corresponding mathematics and chemistry requirements for the "standard" degree program.

Our site visits indicated that thriving departments often involved students in research experiences or required a thesis based on a research or library project. Our survey indicated that a research experience is required by 36% the schools in which the "standard" physics program is a

B.S. in physics, 29% in which it is a B.A. in physics, and 37% of the other bachelor's program. Similarly, a thesis is required in 14% of the B.S. programs, 19% of the B.A. programs, and 17% of the other programs. Perhaps surprisingly, the research and thesis requirements were more common in B.A./B.S. institutions (28% and 13%, respectively) than in Ph.D. institutions (21% and 9%).

Table 4. Physics, mathematics, and chemistry requirements in "standard" degree programs

"Standard" physics degree program	Physics credits required*		Mathematics credits required*		Chemistry credits required*		Number of respondents
	Low	High	Low	High	Low	High	
B.S. in physics	27%	41%	10%	18%	3%	8%	413
B.A. in physics	23%	37%	6%	16%	3%	9%	100
Other bachelor's	22%	40%	7%	17%	4%	9%	32

* Credits are given as a percentage of the total credits required for the degree. Low and high figures represent 10% and 90% range; that is, 10% of the respondents are below the low figure and 10% are above the high figure. Fifty-six percent of the B.A. programs and 22% of the B.S. and other bachelor's programs require no chemistry; the table shows the typical ranges only from the respondents who require chemistry.

The B.S. is by far the most common "standard" degree program: It is identified as the "standard" degree by 71% of institutions in which the bachelor's is the highest degree, 90% of M.S. institutions, and 84% of Ph.D. institutions. The B.A. is the "standard" degree at 23% of the bachelor's institutions, 2% of the M.S. institutions, and 11% of the Ph.D. institutions. Other degrees identified at 6–8% of institutions as the "standard" program include a bachelor's in engineering physics (2% of all institutions) and a bachelor's in applied physics (2% of all institutions).

For the various "standard" tracks, we asked institutions to report the number of credits required for various courses in the physics curriculum. These data are still under analysis to correct for variations in systems of assigning credits.

Table 5 shows the alternative degree tracks offered by institutions with various "standard" physics degree programs. Overall, 84% of departments offer at least one alternative degree track. We asked institutions to specify the number of physics credits required in their various alternative tracks and to list the number of students who completed degrees in the past three years under the alternative tracks. These data proved difficult to analyze, so it may be necessary to do follow-up surveys to selected departments to complete the correlation between the availability of alternative tracks and the number of physics graduates.

Table 5. Fraction of reporting departments offering alternative physics degree tracks for various "standard" programs

Alternative degree track	"Standard" physics degree program			Overall fraction %
	B.S. in physics %	B.A. in physics %	Other bachelor's %	
B.A.	46	n/a	28	39
Physics degree for teachers	30	24	28	29
Specialized degree (e.g., geophysics)	17	13	14	16
Applied physics	16	6	17	14
Engineering physics	15	6	17	13
Combined degree (e.g., physics + math)	11	11	28	12
Astronomy degree	7	11	—	7
Other	20	18	24	20
No alternative track	12	35	24	16
Number of responding departments	368	80	29	477

Questions concerning the availability of minors did not yield any surprising results. Of the reporting departments, 90% offer a minor in either physics (75%), astronomy (1%), or both (14%). As might be expected, the departments with the largest numbers of minors are also those with the largest number of majors. However, the overall numbers are relatively small—only 16% of reporting departments awarded an average of more than two minors per year in physics or astronomy during the past three years. While virtually all physics departments offer minors, there are cultural differences in the extent to which this option is stressed. By way of example, some of the research departments that award very large numbers of physics bachelors annually award no minors in physics. Conversely, some of the smaller bachelors-granting departments award more minors in physics annually than they do bachelors in physics.

Finally, the responders were asked an open-ended question about whether their institutions were planning to add any alternative degree tracks in the near future. Most responders did not answer this question, which we take to suggest that they are not planning to add new tracks. Of those who did respond, the most frequent answers were "no" (156 responses) and "maybe" (29). Other frequently cited responses were engineering (28), teaching (13), applied physics (12), computational physics (11), medical physics (11), and astronomy (9). Other responses included biological physics, materials physics, physics for pre-law, and geophysics.

Recruitment Activities

Question 8 on the survey form asked departments to specify which recruitment activities they pursue. Responses (from 561 departments) were as follows:

Recruiting high school students

Host prospective students and their families in the department	60%
Hold annual departmental open house for students and parents	47%
Recruit high school students likely to major in physics	34%
Faculty and students regularly visit local high schools	24%
Hold summer workshops for high school students	14%
Recruit high school students who are underrepresented minorities	11%

Recruiting enrolled college students

Identify and recruit talented students in service courses	61%
Group potential physics majors in special section of intro course	22%
Offer "introduction to the profession" courses for first year students	15%
Actively recruit transfer students from two-year colleges	11%

The right-hand column adds to more than 100% because departments were asked to indicate all recruiting activities in which they took part.

As might be expected, there is a correlation between the number of recruiting activities and the size of the department. Departments in which the highest degree is the bachelor's reported an average of 2.7 recruiting activities, while Ph.D.-granting departments reported an average of 3.9 recruiting activities. Of the 113 departments reporting 0 or one recruiting activities, 68% had six or fewer faculty; conversely, of the 122 departments reporting five or more recruitment activities, 71% had more than six faculty.

The correlations between recruitment activities and number of degrees awarded are shown in Tables 6 and 7 for departments in which the highest degree is respectively the bachelor's and the Ph.D. Table 6 indicates that, of the bachelor's-granting departments, 50% of those that engage in the fewest recruiting activities (0 or one) awarded fewer than two degrees per year over the three-year period. At the other end of the scale, the correlation is much weaker—of the departments that engaged in the highest number of recruitment activities (four or more), fewer than half exceed the average number of degrees (three/year) awarded by baccalaureate departments.

For the Ph.D.-granting departments (Table 7), the correlation between the number of graduates and the number of recruitment activities is weak at best. Sixty percent of the departments with five or more recruiting activities fail to reach the average of 10 graduates per year that characterizes the Ph.D.-granting institutions, and a third of the departments with the smallest number of recruiting activities exceed the average.

Table 6. Effect of number of recruitment activities on three-year total of physics bachelor's degrees at baccalaureate institutions

Three-year total of bachelor's degrees	Number of recruitment activities					Overall fraction of departments
	≤ 1	2	3	4	≥ 5	
0 to 5	50%	38%	26%	27%	23%	35%
6 to 9	23%	24%	26%	29%	30%	26%
10 to 14	12%	20%	19%	17%	25%	18%
15 or more	15%	18%	28%	27%	21%	21%
Number of responding respondents	100	81	91	52	56	380

Table 7. Effect of number of recruitment activities on three-year total of physics bachelor's degrees at Ph.D.-granting institutions

Three-year total of bachelor's degrees	Number of recruiting activities			Overall fraction of departments
	0 to 2	3 to 4	≥ 5	
0 to 14	27%	28%	28%	28%
15 to 29	42%	37%	32%	36%
30 or more	31%	35%	40%	36%
Number of responding departments	26	57	50	133

Another open-ended question asked departments to specify which of their recruiting activities they considered to be most effective. The most frequent response (which was not given among the original choices) was to assign good teachers to the introductory courses (cited by 29 responders). Other frequently mentioned successful activities included hosting of prospective students, recruiting talented students from the introductory course, high school recruitments, and open houses. Less frequently cited responses (fewer than 10) included special programs or courses, recruitment by the admissions staff, scholarships, telephone contacts, web or email contacts, mailings or brochures, SPS activities, and research opportunities.

Interactions between Faculty and Students

Among the measures of satisfaction most often mentioned by students during our site visits were advising and the informal interactions between students and faculty. Our survey sought to gather additional information on the number and type of these interactions and their correlation with the number of majors.

Overall, most institutions assign several faculty members as undergraduate advisors. A significant number, however, use only a single faculty member (or the department chair) as the undergraduate advisor. Curiously, this distinction correlates inversely with the size of the institution and is mostly independent of the number of majors. Multiple faculty members handle the advising in 75% of bachelor's institutions, 41% of M.S. institutions, and 51% of Ph.D. institutions, while a single faculty member handles the responsibility in 22% of bachelor's institutions, 55% of M.S. institutions, and 41% of Ph.D. institutions. Multiple faculty do the advising at 74% of schools with four or fewer faculty, 80% of schools with 4.1–9 faculty, 46% of schools with 9.1–25 faculty, and 59% of schools with more than 25 faculty; a single faculty member (possibly the department chair) is assigned to the advising in respectively 24%, 18%, 49%, and 27% of schools in these categories. Multiple faculty members do the advising in 67% of schools with an average of fewer than two graduates per year and also in 67% of schools with an average of more than 10 graduates per year, while a single faculty member does the advising in 27% of schools with an average of fewer than two graduates per year and in 23% of schools with an average of more than 10 graduates per year. (A small number of departments use a nonfaculty advisor in the physics department or a university advisor outside of physics.)

Responses regarding the frequency of student-advisor interaction varied from at most once per year to several times per term. Overall 62% of institutions reported that interactions occurred several times per term, with the bachelor's institutions ranging somewhat higher (70%) and the Ph.D. institutions lower (42%). As might be expected, those institutions using multiple faculty advisors were more likely to report several interactions per term (72%) than those using a single faculty member (25%). Unfortunately, during the site visits we did not ask students about the frequency of their interactions with their advisors; it would have been interesting to verify whether the data provided by the department head is "wishful thinking" or reality.

Question 11 of the survey form asked departments to indicate which of a list of activities they engaged in to make their students feel a part of the department. Responses were as follows:

Have an active physics club or SPS chapter	76%
Provide a dedicated undergraduate study room or lounge	74%
Provide building keys to undergraduate majors	52%
Conduct exit interviews with graduating seniors	43%
Assign a faculty mentor to each student	43%
Assign a peer mentor to each student	2%
Other activities	32%

Ph.D.-granting departments tended to run 10–20% above these averages, while bachelor's-granting departments tended to run about 10–20% lower. There was a similar correlation with the size of the department, with 53% of departments having three or fewer faculty engaging in two or fewer of these activities, while 68% of departments with 25 or more faculty engaged in four or more activities.

How do these activities correlate with the department's success in attracting and retaining majors? Tables 8 and 9 display the correlations for bachelor's-only institutions and Ph.D.-granting institutions, respectively. Here the correlations appear to be much stronger than was the case for the correlation between recruitment and number of majors (Tables 6 and 7). It seems clear that departments should focus their efforts on improving these interactions rather than on recruitment. This is consistent with strong anecdotal evidence obtained in conversations with students during the site visits—many students reported switching from engineering or math to physics because the physics department presented a more welcoming and accommodating image.

Table 8. Effect of number of departmental interactions on three-year total of physics bachelor's degrees at baccalaureate institutions

Three-year total of bachelor's degrees	Number of departmental interactions					Overall fraction of departments
	≤ 1	2	3	4	≥ 5	
0 to 5	19%	26%	30%	21%	5%	35%
6 to 9	13%	27%	24%	21%	16%	26%
10 to 14	11%	18%	35%	20%	17%	18%
15 or more	5%	18%	30%	28%	18%	21%
Number of responding respondents	48	85	107	81	48	370

Table 9. Effect of number of departmental interactions on three-year total of physics bachelor's degrees at Ph.D.-granting institutions

Three-year total of bachelor's degrees	Number of departmental interactions			Overall fraction of departments
	0 to 2	3 to 4	≥ 5	
0 to 14	36%	50%	14%	28%
15 to 29	17%	49%	34%	36%
30 or more	15%	52%	33%	36%
Number of responding departments	29	65	37	131

The availability of career information represents another area in which outreach by the department can enhance the student experience, both to attract majors and to help launch imminent graduates toward the next stage of their professional careers. Question 12 of the survey asked departments to list activities undertaken *within the past year* to provide career information to undergraduates. Overall responses were as follows:

Career materials from professional societies	63%
The university career services office	51%
Departmental colloquia by physicists in industry	47%
Alumni visits to the department	45%
Field trips to local industries	25%
Other	26%
None offered	6%

In contrast to the case of departmental interactions, in providing career information the level of activity of bachelor's-only institutions tended to be a bit *above* that of Ph.D.-granting institutions. With the exception of departmental colloquia (which are more common in Ph.D.-granting institutions), the activity level of bachelor's-only institutions in offering career information tended to fall about 10% above these overall averages, while that of the Ph.D.-granting institutions fell about 10% below. The level of activity in this area correlated weakly with the size of the department, with large departments tending to provide a somewhat greater level of career information than smaller departments. It appears from these data that larger, bachelor's-only departments are the most active in providing career information to their majors.

Two open-ended questions were asked in these areas. The first inquired about the department's most successful activities in shaping student attitudes toward the department. The most frequent responses were: informal interactions between faculty and students (cited on 112 forms), undergraduate study room (44), SPS activities (42), research experiences (33), and advising or mentoring (20). The second open-ended question asked about the department's most useful activity in providing career guidance. Responses included: faculty advising (55 citations), alumni visits (31), colloquia (18), research opportunities (15), career service office (10), and career materials from professional societies (9).

Alumni Tracking

Of the 561 responses to the survey, 453 reported answers to question 13, which asked about the career destinations of graduates of the past three years. Table 10 compares the responses from the three types of institutions. It should be noted that these data are given by number of responses and are NOT weighted by the number of graduates.

Table 10. Reported alumni destinations (as percent of responses) by highest degree awarded by institution.

Alumni destination	Highest institutional degree			Overall %
	B.S./B.A.	M.S.	Ph.D.	
Graduate school in physics	31	38	43	35
Other graduate school	15	7	12	14
Continue in 3/2 engineer. prog.	11	6	1	8
Employed in technical field	22	28	21	22
High school teaching	8	7	4	7
Employed in nontechnical field	3	2	3	3
Active military	2	1	2	2
Other	2	3	2	2
Don't know	7	8	13	9
Number of responding departs.	300	42	111	453

The data in Table 10 agree reasonably well with the data collected in the AIP Survey Research Center's annual survey of physics departments, which in recent years indicates that 32% of graduates enter graduate programs in physics, 20% enter graduate programs in other fields, and 48% enter the workforce.

Question 14 asked departments to identify activities they used to keep in contact with their alumni. With one exception, responses varied little among B.A./B.S., M.S., and Ph.D. institutions. Overall responses were:

Updates from past students by email and phone	51%
Mailing or email addresses from students at graduation	46%
Information on employment or graduate school plans at graduation	45%
Mailing list for departmental newsletter	26%
Surveys of alumni	24%
Other	4%
None of the above	32%

The one case in which there was a significant variation among types of institutions was in that of the departmental newsletter, for which 45% of Ph.D. institutions indicated that choice but only 20% of B.A./B.S. institutions.

Curricular Reform

More than 60% of the reporting departments (342 out of 561) affirmed that they had made "significant" changes in curriculum in the past several years. The overall responses to specific areas of change were as follows: upper-division courses—71%; calculus-based introductory courses—70%; general education courses—56%; introductory courses for majors—51%; algebra-based introductory courses—42%. Among the three types of institutions (B.S./B.A., M.S., Ph.D.), B.S./B.A. institutions were more likely than Ph.D. institutions to have made changes in their upper-division courses (76% vs. 60%) and general education courses (60% vs. 44%), while Ph.D. institutions were more likely than B.S./B.A. institutions to have made changes in calculus-based introductory courses (75% vs. 69%) and introductory courses for majors (64% vs. 47%).

It was more common for departments to report reforms in both content and pedagogy than in either area alone; the fractions of the reforms involving both content and pedagogy ranged from about 60% in general education courses, to 50% in introductory courses for majors, to 40% in calculus-based introductory courses and upper-division courses, and to 30% in the algebra-based introductory courses. Changes involving content only were most common in the upper-division courses (40%) and rather rare for introductory courses (10–20%). These data are not surprising, in that departments often introduce new courses for majors and rarely alter the traditional curriculum of the introductory courses. Changes in pedagogy more commonly occurred at the introductory level (80–90% of the changes involved pedagogy, either alone or in concert with changes in content) than at the advanced level (56%). There was relatively little systematic variation across institutional types in these data.

These results paint an encouraging profile of the state of undergraduate physics education and suggest a greater widespread receptiveness for curricular change than had been anticipated before the survey (although we must admit that we left it to the departments to determine for themselves what is a "significant" curricular reform).

The most common funding mechanism for curricular reforms was by far the internal reallocation of resources within the department. Overall 63% of those who reported curricular changes indicated this as a source of funding. (Responders were free to indicate several sources.) Among institutional types, these responses ranged from 57% at the B.S./B.A. institutions to 76% at the Ph.D. institutions, presumably because the latter often have more resources to reallocate. Other sources of funding were university or endowment funds from outside the department (27%), grants from NSF or other federal agency (22%), grants from private foundations (9% overall, but more common at B.S./B.A. institutions than at Ph.D. institutions), funds or equipment from business or industry (4%). Except as noted above, there was little variation across institutional types in these responses. It should be especially noted that the share from the federal agencies remains relatively flat across institutional types, indicating that the large research institutions are clearly not dominant in the quest for federal funding for curricular reform (although the dollar-weighted funding averages might be skewed in that direction).

We also asked three open-ended questions about the curricular reforms. Regarding the source of the motivation of the reforms, by far the most common response was energetic individual faculty members (112 responses). The next most common reform motivation was the desire to improve courses (41 responses), followed by college-wide initiatives (26), responses to external reviews (18), responses to student feedback (18), the desire to recruit more majors (16), internal reviews (15), and the development of new programs (11). "Negative" motivations (threats from the dean, reduction in faculty size) were cited only rarely (respectively, eight and five responses). The impact of physics education research was cited only three times, although that may have been an indirect motivation in driving the "energetic individual faculty members" to undertake the reform efforts.

The most common response to our inquiry about what the reforms were intended to accomplish was attract more majors (71), although content-related issues (better understanding of topic, better preparation for graduate school, better courses, better preparation for work force) drew significant responses as well (respectively 48, 47, 40, 25).

It appears that departments undertake these changes with only vague ideas of assessment. Our query regarding indicators of success drew "none" as its most common response (59), although student attitudes (43), increased number of majors (40), and increased enrollments (21) were also cited. Only 39 responses mentioned a formal assessment mechanism such as the Force Concept Inventory exam or the Graduate Record Examination.

Overall Evaluations of Undergraduate Physics Programs

Our survey concluded with open-ended questions about the undergraduate program's greatest strengths and greatest needs or challenges. Individualized faculty attention to students was the overwhelming choice for the greatest strength (203 responses). Next most often cited responses were research opportunities (89), excellent curriculum (79), and quality of faculty (70). Few departments cited flexible major programs (17) or an active SPS chapter (3) among their strengths; this is rather surprising in view of the survey results showing that 84% of departments offer alternative tracks to the major and 76% of departments regarded their SPS chapter as a major factor in promoting a welcoming attitude toward students. Only 15 departments cited their excellent equipment as one of their greatest strengths.

By far the greatest challenge seen by most department heads is the need for more students (204 responses). The need for resources is next most commonly cited, including more faculty (73), improved lab equipment and space (41), increased funding (39), increased administrative support (17), improved research opportunities (20), and better facilities and more space (13). Improved quality is also a significant need, including courses (38), faculty quality (17), better student preparation in math (20), and better students (9). Sadly, increased minority representation drew only five responses.

Chapter 6.
Connections, Lessons, and Other Issues

In the previous sections of this report, we have laid out our analysis of what makes a thriving undergraduate physics program. That analysis has been backed up with evidence and examples. In this section, we take a somewhat more subjective approach to look at connections between our analysis and work done by others. We also write about some new questions that have arisen as part of our study and implications of those questions for the future of physics.

Undergraduate Mathematics Site Visits

After this study was well under way, we learned about a similar effort in undergraduate mathematics education. (That in itself says a lot about how poorly the scientific and mathematics communities communicate the results of such studies.) In the early 1990s, the Mathematical Association of America, with funding from the National Science Foundation, carried out a series of 10 site visits to undergraduate mathematics programs. Although the mathematics site visit group used a methodology somewhat different from that used by SPIN-UP, the conclusions expressed in the report *Models that Work: Case Studies in Effective Undergraduate Mathematics Programs* [Tucker, 1995] are quite consistent with the results we have found from the SPIN-UP site visits. The MAA group selected departments based on evidence of effective practices excelling in

- attracting and training large numbers of mathematics majors, or

- preparing students to pursue advanced study in mathematics, or

- preparing future school mathematics teachers, or

- attracting and training underrepresented groups in mathematics.

On p. vii the report states, "The site visits revealed that there is no single key to a successful undergraduate program in mathematics. Almost any approach can be made to work in almost any institutional context if a substantial number of the mathematical faculty care deeply about undergraduate education, create an atmosphere among faculty and students that the study of mathematics is important and rewarding, and maintain close interactions with their students." This finding agrees with what SPIN-UP found in the 21 physics site visit departments.

The MAA report goes on to delineate the common features of effective programs (p. 3) including three states of mind that underlie faculty attitudes in effective programs:

- respecting students, and in particular, teaching for the students one has, not the students one wished one had.

- caring about the students' academic and general welfare.

- enjoying one's career as a collegiate educator.

"A common theme of effective programs is the existence of a variety of mechanisms for interactions between faculty and students outside of class, both in one-on-one settings and in social groups." (p. 3) The effective departments also exhibited:

- a curriculum geared toward the needs of the students not the values of the faculty

- an interest in using a variety of pedagogical and learning approaches.

The site visit committee identified four major components of efforts to reform mathematics education: (p. 32)

a. Assessing the goals of the current program and aligning them with the needs of the students.

b. Building support for innovation that engages the faculty. (Some efforts start with broad support within the department, others are initiated by a few energetic individuals.)

c. Initiating the process of change and experimentation. Continuing experimentation was the hallmark of most of the institutions visited in this report, even though they already had successful programs.

d. Developing an environment of faculty involvement in the welfare, academic, and otherwise, of their students.

All of these statements closely mirror the results found in the SPIN-UP site visits. Our conclusion is that the features of effective undergraduate programs apply to all the science, technology, engineering, and mathematics disciplines.

Revitalizing Undergraduate Science Education

Sheila Tobias has been studying science and mathematics education for more than a decade. Her book *Revitalizing Undergraduate Science Education* (Research Corporation, Tucson, AZ, 1992) reports on a series of case studies in undergraduate chemistry, physics, and mathematics programs that "work." Tobias's conclusions neatly parallel those in this SPIN-UP report. Here are some selected conclusions:

> "…first, change is not implemented by experts, but originates in local commitment and reallocation of resources at the midlevel of management—in the case of colleges and universities, the department." (p. 158)

> "A hallmark of effective programs is that the process of reform is all-engaging. Ideas are solicited from faculty and implemented locally by the department." (p. 158)

> "The model for science education reform is not an experimental model, not even a research model, but a *process model* that focuses attention continuously on every aspect of the teaching-learning enterprise, locally and in depth…. In programs that work, faculty members pay continuous attention to 'what we teach, who we teach, and how we teach.'" (p. 160)

Thriving in the Business World

Parallels to our report's conclusions are also found in the vast business literature on managing change and building thriving companies. As one recent example, we cite Jim Collins's *From Good to Great* (HarperCollins, New York, 2001), which analyzes a number of companies that have successfully made the transition from being "good" to being "great." Collins and his research team identified a number of characteristics shared by these companies and missing in those companies that failed to make the transition. Again, a few selections show the parallels:

> "All good-to-great companies began the process of finding a path to greatness by confronting the brutal facts of their current situation.... The good-to-great companies faced just as much adversity as the comparison companies, but responded to that adversity differently. They hit the realities of their situation head on." (p. 88)

> "Good-to-great transformations often look like dramatic, revolutionary events to those observing from the outside, but they feel like organic, cumulative processes to people on the inside. The confusion of end outcomes (dramatic results) with process (organic and cumulative) skews our perception of what really works over the long haul. No matter how dramatic the end result, the good-to-great transformations never happened in one fell swoop. There was no single defining action, no grand program, no one killer innovation, no solitary lucky break, no miracle moment." (p. 186)

> "Level 5 leaders [leaders of those companies that have made the transition] are ambitious for the company and what it stands for; they have a sense of purpose beyond their own success." (p. 198)

Underrepresented Groups and the Issue of Diversity

It is a well known but still unsettling fact that women and minorities are distinguished by their lack of presence in the STEM disciplines, particularly in the physical sciences, mathematics and engineering. The National Science Foundation's *Science and Engineering Indicators 2002* gives the detailed statistics. Participation is increasing, but much more slowly than everyone would like. There is much speculation about this lack of participation, and we shall not rehearse those speculations here. The SPIN-UP site visits did uncover one surprise: We had anticipated that thriving departments, which managed to recruit many more students than the national average and which were well regarded within their institutions, would have substantial success in bringing women and minority students into physics. We found that most of the site visit departments did in fact do a bit better than the national average in attracting underrepresented students, but not a lot better. This finding was a surprise to all of the site visit teams because the folklore amongst those who are actively working to increase the participation of underrepresented groups in STEM is that active, supportive programs will be much more successful in attracting women and minority students. Our conclusion is that these conditions may be necessary, but they are not sufficient.

The MAA report [Tucker, 1995] comes to a similar conclusion:

"Unfortunately, the four-year mathematics programs visited in this project had negligible numbers of Black and Latino mathematics majors, except of course for Southern University and Spelman College, which are historically Black institutions..... The programs that draw large numbers of other types of students apparently need to do something different and special for attracting and retaining this group in mathematics." (page 30)

Faced with these surprising results, the Task Force decided to invite about a dozen representatives of the National Society of Black Physicists and the National Society of Hispanic Physicists to its December 2002 meeting to discuss this issue and to explore possible studies the Task Force might undertake to understand this critical issue more fully. The participants at the meeting concluded that nearly all the general factors that seem to be important for attracting and retaining underrepresented groups in physics are the same factors that attract and retain "traditional" physics students. Nevertheless, these factors do not guarantee that a particular department will attract more students from underrepresented groups. The Task Force plans to use its site visit methodology to study a number of undergraduate physics programs that in fact do serve larger numbers of minority students. Many of these, of course, will be historically black colleges and universities. We also plan to make use of the results of a recent study (Barbara Whitten, Colorado College) that focused on physics departments that have a large fraction of women physics majors.

We argue that increasing the presence of underrepresented groups in physics is important on two counts: First, it is just the right thing to do. Everyone should have the opportunity to experience the joys (and frustrations) of science and to contribute to the betterment of humankind through science. Second, in the 21st century the population subgroup that has historically dominated science (white males) will shrink both in absolute numbers and as a fraction of the U.S. population. It is difficult, if not impossible, to say precisely how many scientists, engineers, and mathematicians the United States needs, but it is safe to say that number will not decrease from our current needs. We do know that with more scientists and with more diverse backgrounds represented, science is likely to advance more rapidly than it would otherwise. Simply from the perspective of maintaining a vibrant scientific and technological workforce to maintain our economy, our security, and the infrastructure needed to improve the health and environment for all people, we will need to tap the full spectrum of the nation's talent for the next generation of scientists, mathematicians, and engineers.

As a result of the December meeting, the Task Force will carry out two further site visits, one to an historically black college or university and the other to a hispanic-serving institution. The results of those site visits will be reported elsewhere. Some Task Force members and some of the representatives at the December meeting are writing a report on the meeting along with articles to be submitted to various physics publications.

Two-year Colleges

The SPIN-UP study focused on physics departments that offer at least the bachelor's degree in physics. Nationwide, however, many students study physics in two-year colleges, some in preparation for transfer to a four-year institution, some as part of their technology training for a two-year associate's degree. A study carried out by the American Institute of Physics [Neuschatz, et al., 1998] indicates that about half of the nation's pre-service teachers take their science courses at two-year colleges. Two-year colleges play an important role in undergraduate science education. But, the academic organization of two-year colleges is substantially different from that of four-year colleges and universities. Only the largest of the two-year colleges have physics departments. Most employ one or two physics faculty as part of a department of natural sciences. In order to understand the characteristics that make up a thriving physics program at a two-year college, Task Force member Tom O'Kuma (Lee College) and Mary Beth Monroe (Southwest Texas Junior College) developed project "SPIN-UP Two-Year Colleges" that has been funded by the National Science Foundation. Employing a site visit protocol similar to that used for SPIN-UP, the two-year college study will sponsor visits to 10 or so two-year colleges and produce a report analyzing the results of those site visits. The report is expected to be finished during the fall of 2003.

Teacher Preparation

Both the general public and those within the scientific community have been calling for improved preparation for K–12 teachers in mathematics and science. Much of this responsibility must fall on the shoulders of undergraduate science and mathematics programs. We were disappointed to find that most of our site visit departments were not actively engaged in pre-service teaching preparation either at the K–8 or at the high school level. Those that did have programs for teacher preparation were serving relatively few students. Although most of the departments acknowledged that they should be doing more, they cited difficulties working with the School of Education (or its equivalent) and the lack of student interest. Even when pre-service courses were offered, few students (and even fewer physics majors) took them.

The physics professional organizations have recently urged physics departments to take a more active role in both pre-service and in-service work with teachers. With generous funding from the National Science Foundation and the Fund for Post-Secondary Education, AAPT, APS, and AIP have launched the Physics Teacher Education Coalition to develop models of effective programs for pre-service teacher work within physics departments. More information on this program can be found at *http://www.phystec.org*. In addition, the report of a University of Nebraska–Lincoln conference on teacher preparation in physics departments [Buck, Hehn and Leslie-Pelecky, 2000] contains several articles describing what physics departments can do to aid pre-service teachers.

Future Directions

1. The Task Force is now in the process of preparing proposals for activities that will build on the SPIN-UP results. The aim is to work with physics departments that want to "revitalize" their undergraduate physics programs. A trial workshop held at the PKAL Summer Institute June 2002 drew about 65 participants from 20 or so physics departments. In the spring of 2003, the Task Force will carry out a trial "consulting visit" to aid a department

planning the revitalization of its undergraduate program. A proposal for more workshops and consulting visits to reach, say, 150 physics departments, would have a major impact on undergraduate physics in the United States

2. As we mentioned previously, the Task Force will also explore the issues of diversity in physics with the aim of promoting some concrete activities.

3. Plans are under way for a conference on algebra-based introductory physics courses to complement the conference on calculus-based introductory physics.

4. The AAPT Committee on Undergraduate Physics, under the leadership of Steve Turley of Brigham Young University, is undertaking a revision of AAPT's *Guide to Undergraduate Physics Programs*. The results of SPIN-UP will guide these revisions. Physics departments planning new initiatives or preparing for departmental reviews often use this booklet.

5. The Task Force has begun to work with leaders in other STEM disciplines who are focusing on undergraduate education. For example, a member of the leadership of ProjectNEXT, Tom Rishel, attended the 2000 New Physics Faculty Workshop, and we hope to send a member of the Task Force to their next conference. In June 2002, Hilborn and Hehn met with geologists Cathy Manduca (Carleton College), David Mog (Montana State University), and Heather MacDonald (William and Mary) to discuss possible extension of the physics site visit techniques to geology departments and other possible areas of collaboration. There has also been some contact with the American Chemical Society's Committee on Professional Training for a possible joint study of diversity issues.

Final Words of Wisdom and Encouragement

The Task Force on Undergraduate Physics is committed to the improvement of undergraduate physics because undergraduate physics plays an absolutely crucial role in educating the next generation of scientists and engineers, the next generation of K–12 teachers, and the future leaders of our society. We believe that the conclusions drawn from our analysis of the site visits and the general survey provide a blueprint for what is needed to build a thriving undergraduate physics program. The blueprint, however, must be adapted to fit each department's local "zoning regulations" (the students each department serves, the faculty and physical resources available, and the mission of the home institution). But we are convinced that with sustained efforts every physics department can have a thriving program in which students are challenged and supported in their many career and intellectual goals and faculty find great satisfaction in approaching undergraduate physics as a scholarly enterprise worthy of the problem-solving and critical-thinking skills that sustain them as researchers.

References

J.D. Bransford, A.L. Brown and R.R. Cocking, eds., *How People Learn, Brain, Mind, Experience, and School* (National Academy Press, Washington, DC, 1999).

G. Buck, J. Hehn and D. Leslie-Pelecky, eds., *The Role of Physics Departments in Preparing K–12 Teachers* (American Institute of Physics, College Park, MD, 2000).

L.A. Coleman, D.F. Holcomb and J. S. Rigden, "The Introductory University Physics Project 1987–1995: What has it accomplished?," *Am. J. Phys.* **66,** 124–137 (1998).

M.D. George, S. Bragg, J. Alfredo G. de los Santos, D.D. Denton, P. Gerber, M.M. Lindquist, J.M. Rosser, D.A. Sanchez and C. Meyers, *Shaping the Future: New Expectations for Undergraduate Education in Science, Mathematics, Engineering, and Technology NSF 96-139* (National Science Foundation, Washington, DC, 1996).

E. Mazur, *Peer Instruction* (Prentice Hall, Upper Saddle River, New Jersey, 1997).

M. Neuschatz, G. Blake, J. Friesner and M. McFarling, *Physics in the Two-Year Colleges* (American Institute of Physics, College Park, MD, 1998).

G.M. Novak, E.T. Patterson, A.D. Gavrin and W. Christian, *Just-In-Time Teaching: Blending Active Learning and Web Technology* (Prentice-Hall, Upper Saddle River, NJ, 1999).

D. Sokoloff and R. Thornton, "Using Interactive Lecture Demonstrations to Create an Active Learning Environment," *Phys. Teach.* **35**, 340–346 (1997).

R. Thornton and D. Sokoloff, "Assessing student learning of Newton's laws: The force and Motion Conceptual Evaluation and the Evaluation of Active Learning Laboratory and Lecture Curricula," *Am. J. Phys.* **66,** 338–352 (1998).

A.C. Tucker, *Models that Work: Case Studies in Effective Undergraduate Mathematics Programs* (The Mathematical Association of America, Washington, DC, 1995).

Appendices

Appendix I.
Physics Education Resources

(compiled by Jose P. Mestre)

In this section we present a series of brief descriptions of recently developed materials for undergraduate physics. Each description was prepared by one of the authors of the materials. Each author also provided a series of written references and URLs for further information. The Task Force does not intend to endorse any of these resources over others that are not included in this appendix. These are the ones for which we received responses to a widely distributed solicitation within the physics community.

A. Physlets

Wolfgang Christian, Davidson College

The Physlet project is a synergy of curriculum development, computational physics, and physics education research. This project distributes a wide variety of class-tested interactive materials based on Java applets. Physlets employ a scripting language (JavaScript) to customize applets embedded within HTML pages, thereby allowing one applet to be used in many different contexts. This modular object-oriented software design enables Physlet adopters to easily author and customize their own interactive problems.

References

W. Christian and M. Belloni, *Physlets* (Prentice Hall, Upper Saddle River, NJ, 2001).

G. Novak , E. T. Patterson, A. Gavrin, and W. Christian, *Just In Time Teaching* (Prentice Hall, Upper Saddle River, NJ, 1999).

Website: *http://webphysics.davidson.edu/Applets/Applets.html*

B. Scale-Up

Robert Beichner, North Carolina State University

SCALE-UP stands for "Student-Centered Activities for Large Enrollment Undergraduate Programs." We are adapting research-based pedagogies like collaborative, interactive learning so that they can be used in large-enrollment courses. This is done in a redesigned classroom environment of round tables and laptop computers where special classroom management techniques are utilized. Students are assigned to collaborative groups and spend most in-class time working on "tangible" (hands-on) and "ponderable" ("minds-on") activities. The instructor and assistant(s) circulate and engage in Socratic dialogs with the students.

References

The precursor to SCALE-UP is described in R. Beichner, L. Bernold, E. Burniston, P. Dail, R. Felder, J. Gastineau, M. Gjertsen, and J. Risley, "Case study of the physics component of an integrated curriculum," *Am. J. Phys.* (Phys. Ed. Res. Supplement) **67**, S16S24 (1999).

http://www.ncsu.edu/per/Articles/04IMPEC_AJP.pdf

Also see:

Robert J. Beichner, Jeffrey M. Saul, Rhett J. Allain, Duane L. Deardorff, David S. Abbott, "Introduction to SCALE UP: Student-Centered Activities for Large Enrollment University Physics," Proceedings of the 2000 Annual meeting of the American Society for Engineering Education (2000).

http://www.ncsu.edu/per/Articles/01ASEE_paper_S-UP.pdf

J. Saul, D. Deardorff, D. Abbott, R. Allain, and R. Beichner, "Evaluating introductory physics classes in light of ABET criteria: An Example of SCALE-UP Project," Proceedings of the 2000 Annual meeting of the American Society for Engineering Education (2000).
http://www.ncsu.edu/per/Articles/02ASEE2000_S-UP_Eval.pdf

R. Beichner, "Student-Centered Activities for Large Enrollment University Physics (SCALE-UP)," Proceedings of the Sigma Xi Forum "Reshaping Undergraduate Science and Engineering Education: Tools for Better Learning," Minneapolis, MN (2000).
ftp://ftp.ncsu.edu/pub/ncsu/beichner/RB/SigmaXi.pdf

Scale-Up Website*: http://www.ncsu.edu/per/scaleup.htm*

C. Workshop Physics

Priscilla Laws, Dickinson College

Workshop Physics is a calculus-based introductory curriculum designed to help students understand the basis of knowledge in physics through the interplay between observations, experiments, definitions, mathematical description, and the construction of theories. Instead of attending separate lecture and lab sessions, they attend three 2-hour-long sessions each week to predict, observe, experiment, and use a powerful set of computer tools to develop graphical and mathematical models of phenomena. The curriculum is embodied in a 28 Unit Activity Guide published by John Wiley & Sons.

References

P.W. Laws, "Calculus-Based Physics Without Lectures," *Physics Today* **44** (12), 24–31 (December 1991).

P.W. Laws, *Workshop Physics Activity Guide* (John Wiley & Sons, Inc., New York, NY, 1997).

P.W. Laws, "Millikan Lecture 1996: Promoting active learning based on physics education research in introductory physics courses," *Am. J. Phys.* **65,** 14–21 (1997).

P.W. Laws, "A New Order for Mechanics," Proceedings of the Conference on the Introductory Physics Course, P.W. Laws and J. Wilson, eds. (John Wiley & Sons, New York, NY, 1997), pp. 125–136.

J.M. Saul, "An Evaluation of the Workshop Physics Dissemination Project" (U. of Maryland, 1998).

Workshop Physics Website: *http://physics.dickinson.edu*

D. Investigative Science Learning Environment (ISLE)

Eugenia Etkina, Rutgers University

Alan Van Heuvelen, Ohio State University and Rutgers University

In ISLE students use the processes of science to construct and apply knowledge. They observe simple experiments, make qualitative explanations and develop quantitative laws, and devise experiments to test and, if needed, revise the laws. The laws and models are applied for useful purposes to real-world applications. These processes of science investigation are integrated with the results of research about learning—active student participation, multiple representations of processes, and multiple exposures to concepts.

References

E. Etkina and A. Van Heuvelen, "Investigative Science Learning Environment: Using the processes of science and cognitive strategies to learn physics," Proceedings of the 2001 Physics Education Research Conference, Rochester, NY, pp. 17–21 (2001).

Websites:
http://www-rci.rutgers.edu/%7Eetkina/isle.htm
http://www.pt3.gse.rutgers.edu/physics/frontp.html

E. ALPS and ActivPhysics (Active Learning in Large and Small Classes)

Alan Van Heuvelen, Ohio State University and Rutgers University

The *Active Learning Problem Sheets* (the *ALPS Kits*) are paper-and-pencil activities that help students participate in learning in lectures and recitations. The kits include qualitative questions, multiple-representation activities, and problems done in a multiple-representation format. *ActivPhysics* is a comprehensive multimedia product that has similar activities as in the ALPS Kits with the added advantage of providing simultaneous simulated processes and dynamic representations of these processes.

References

Alan Van Heuvelen, "Millikan Lecture: The Workplace, Student Minds, and Physics Learning Systems," *Am. J. Phys.* **69,** 1139–1146 (2001).

Alan Van Heuvelen, Active Learning Problem Sheets: Mechanics and Electricity and Magnetism (Hadyn-McNeil, Plymouth, MI,1990).

Alan Van Heuvelen and P. D'Alessandris, *ActivePhysics I and II* (Addison-Wesley-Longman, Palo Alto, CA, 1998).

F. Matter and Interactions, Electric and Magnetic Fields

Bruce Sherwood and Ruth Chabay, North Carolina State University

The two-volume introductory calculus-based college physics textbook *Matter & Interactions* by Ruth Chabay and Bruce Sherwood (Wiley 2002) emphasizes the atomic nature of matter with macro-micro connections and engages students in modeling complex phenomena, including computer modeling. Analyses start from a small number of fundamental principles. Mechanics and thermal physics are treated as a single unified subject, as are electrostatics and circuits. The intent is to make introductory physics reflect the contemporary physics enterprise.

References

R.W. Chabay and B.A. Sherwood, "Bringing atoms into first-year physics," *Am. J. Phys.* **67,** 1045–1050 (1999).

http://www.wiley.com/college/chabay for the textbook, with a link from there to our own public website with additional materials, including free educational software. See *http://vpython.org* for the 3D programming environment developed for use with our curriculum.

G. Teaching Physics Through Cooperative Group Problem Solving

Ken Heller and Patricia Heller, University of Minnesota

Students solve Context-Rich quantitative problems that emphasize making expert-like decisions based on physics concepts. Student support includes teaching a general problem-solving framework and coaching using cooperative groups. This approach follows the Cognitive Apprenticeship paradigm of modeling, coaching, and fading. The modeling of desired problem-solving behavior is in lectures and in written problem solutions while coaching occurs in discussion sections and laboratories where the students work Context-Rich problems in cooperative groups.

References

P. Heller, R. Keith, and S. Anderson, "Teaching problem solving through cooperative grouping. Part 1: Groups versus individual problem solving," *Am. J. Phys.* **60,** 627–636 (1992).

P. Heller and M. Hollbaugh, "Teaching problem solving through cooperative grouping. Part 2: Designing problems and structuring groups," *Am. J. Phys.* **60,** 637–645 (1992).

P. Heller, T. Foster, and K. Heller, "Cooperative group problem solving laboratories for introductory classes," in E. F. Redish and J. S. Rigden, eds. *The Changing Role of Physics Departments in Modern Universities: Proceedings of International Conference on Undergraduate Physics Education* (American Institute of Physics, Woodbury, NY, 1996).

Website: *http://www.physics.umn.edu/groups/physed/*

H. Peer Instruction

Eric Mazur and Catherine Crouch, Harvard University

Peer Instruction engages students in class by asking questions that require each student to apply the core concepts being presented, and then to explain those concepts to their fellow students. Class consists of short lecture segments interspersed with a related conceptual question, called a ConcepTest, which probes students' understanding of the ideas just presented. Students formulate individual answers, then discuss their answers with others sitting around them for two to four minutes. Finally, the instructor calls an end to the discussion, explains the answer, and moves on to the next topic.

References

Catherine H. Crouch and Eric Mazur, "Peer Instruction: Ten Years of Experience and Results," *Am. J. Phys.* **69,** 970–977 (2001).

Adam P. Fagen, Catherine H. Crouch, and Eric Mazur, "Peer Instruction: Results from a Range of Classrooms," *Phys. Teach.* **40,** 206–209 (2002).

Eric Mazur, *Peer Instruction: A User's Manual* (Prentice Hall, Upper Saddle River, NJ, 1997). Website: *http://galileo.harvard.edu/lgm/pi* (Note that this website requires free registration, so on your first visit, you get bounced to a login page which provides links to a registration area).

I. Just-in-Time Teaching (JiTT)

Evelyn Patterson, Air Force Academy

Just-in-Time Teaching (JiTT) is a teaching and learning strategy that exploits interaction between web-based study and an active learner classroom. Students respond electronically to carefully constructed web-based assignments due shortly before class, and the instructor reads the student submissions "just-in-time" to adjust the lesson to suit the students' needs. The heart of JiTT is the "feedback loop" formed by the students' outside-of-class preparation that fundamentally affects what happens during the subsequent in-class time together.

References

G. Novak, E. Patterson, A. Gavrin, and W. Christian, *Just-in-Time Teaching: Blending Active Learning with Web Technology* (Prentice Hall, Upper Saddle River, NJ, 1999).

Webreport: *http://www.pkal.org/pubs/Rothman.pdf*

"Then, Now, and in the Next Decade: A Commentary on Strengthening Undergraduate Science, Math, Engineering and Technology Education" publication which features JiTT on p. 18.

Webreport: *http://a-s.clayton.edu/henry/JiTT.htm*

Gregor Novak and Joan Middendorf, "Just-in-Time Teaching: Using Web Technology To Increase Student Learning," ISETA *Connexions* Newsletter **14** (1), (Spring 2002).

http://webphysics.iupui.edu/JiTT/CATE2000.doc

Gregor Novak and Evelyn Patterson, "The Best of Both Worlds: WWW Enhanced In-Class Instruction," in the Proceedings of the IASTED "Computers and Advanced Technology in Education" [CATE] 2000 International Conference, May 24–27, 2000.

JiTT website: *http://www.jitt.org or http://webphysics.iupui.edu/jitt/jitt.html*

J. Tutorials in Introductory Physics

Lillian McDermott, University of Washington

The Physics Department at the University of Washington has implemented a system of tutorials throughout the introductory calculus-based course. Beginning in 1991 with one lecture section in the mechanics portion of the course, weekly tutorials subsequently became an integral part of the entire first-year sequence, including the honors section. The instructional materials that are used in the 50-minute small sections have been published in *Tutorials in Introductory Physics*. The development of the tutorials has been guided by ongoing research on the learning and teaching of physics and includes continuous assessment through pretests and post-tests. Rigorous T.A. preparation and examinations that include questions on the content in the tutorials are essential for effective adoption. Although there is no direct evidence that the tutorials or the associated T.A. preparation are responsible, their inclusion in the department's instructional program correlates with a rise in the number of graduating physics majors to more than 50 in 2002.

The tutorials comprise an integrated system of pre-tests, worksheets, homework assignments, and post-tests. The tutorial sequence begins with a pre-test that helps students identify what they do and not understand about the material and what they are expected to learn in the upcoming tutorial. The pre-tests also inform the instructors about the level of student understanding. The worksheets, which consist of carefully sequenced tasks and questions, provide the structure for the tutorial sessions. Students work together in small groups, constructing answers for themselves through discussions with one another and with the tutorial instructors. The tutorial instructors do not lecture but ask questions designed to help students find their own answers. The tutorial homework reinforces and extends what is covered in the worksheets.

Post-test results, published in a number of articles, show a significant improvement in student understanding as a result of the tutorials. Furthermore, there has been no decrease in the ability of students to solve standard quantitative problems even though less time is spent in practice on problem solving. Results from pilot sites, ranging from two-year colleges to research universities, demonstrate that the tutorials work equally well in calculus-based and algebra-based courses.

Supported in part by the National Science Foundation, the development of Tutorials in Introductory Physics has been a collaborative effort by all members of the Physics Education Group at the University of Washington, past and present, with contributions by colleagues at other institutions. Leadership in the ongoing development of the tutorials is provided by Lillian C. McDermott, Peter S. Shaffer, and Paula R. L. Heron.

References

Lillian C. McDermott, Peter S. Shaffer, and the Physics Education Group at the University of Washington, *Tutorials in Introductory Physics* (Prentice Hall, Upper Saddle River, NY, Preliminary Edition 1998, First Edition 2002, and Instructor's Guide 2003).

For articles that discuss the motivation for the tutorials and provide an overall description, see:

L.C. McDermott, Millikan Award 1990, "What we teach and what is learned: Closing the gap," *Am. J. Phys.* **59** (4) 301 (1991).

L.C. McDermott, Guest Comment: "How we teach and how students learn—A mismatch?" *Am. J. Phys.* **61** (4) 295 (1993).

L.C. McDermott, Response for the 2001 Oersted Medal, "Physics education research: The key to student learning," *Am. J. Phys.* **69** (11) 1127–1137 (2001).

For articles that illustrate the research that guided the tutorials and describe some specific instructional strategies, see, for example:

L.C. McDermott and P.S. Shaffer, "Research as a guide for curriculum development: An example from introductory electricity, Part I: Investigation of student understanding," *Am. J. Phys.* **60** (11) 994 (1992); "Part II: Design of instructional strategies," *ibid.* **60** (11) 1003 (1992); Printer's erratum to Part I, *ibid.* **61** (81) (1993).

B.S. Ambrose, P.S. Shaffer, R.N. Steinberg, and L.C. McDermott, "An investigation of student understanding of single-slit diffraction and double-slit interference," *Am. J. Phys.* **67** (2) 146 (1999).

K. Wosilait, P.R.L. Heron, P.S. Shaffer, and L.C. McDermott, "Addressing student difficulties in applying a wave model to the interference and diffraction of light," *Am. J. (Phys. Ed. Res. Supplement)* **67** (7) S5 (1999).

The group's URL is *http://www.phys.washington.edu/groups/peg/*.

K. Classroom Communication Systems: Transforming Large Passive Lecture Classes into Interactive Learning Environments

Bill Gerace and Jose Mestre, University of Massachusetts–Amherst

How does one provide a pedagogically sound experience for students enrolled in introductory science classes at large universities, which are commonly taught in large lecture formats numbering from 100–400 students? An emerging technology, classroom communication systems (CCSs), has the potential to transform the way we teach science in large lecture settings. CCSs can serve as catalysts for creating a more interactive, student-centered classroom in the lecture hall, thereby allowing students to become more actively involved in constructing and using knowledge. CCSs not only make it easier to engage students in learning activities during lecture but also enhance the communication among students, and between the students and the instructor. This enhanced communication assists the students and the instructor in assessing understanding during class time, and affords the instructor the opportunity to devise instructional interventions that target students' needs as they arise. In short, CCSs greatly facilitate the instructor's ability to provide an active learning experience for students, to provide feedback to students on their learning, to accommodate different learning styles, to make students' thinking visible, and to provide instruction tailored to students' learning needs—all desirable instructional strategies based on learning principles described in a new report from the National Research Council titled *How People Learn: Brain, Mind, Experience and School.*

Classtalk and Personal Response System (PRS) are two CCSs being used extensively at UMass–Amherst. They are both a combination of software and hardware that permit the presentation of questions for small-group consideration, as well as the collection of answers and the class-wide display of a histogram of student answers. The display of the histogram is the springboard for a class-wide discussion of the ideas and methods used to analyze situations and solve problems. The time devoted to lecturing is decreased, while the time students devote to developing and refining their conceptual understanding is increased. The instructor's role, therefore, more closely resembles that of a coach than a dispenser of information.

CCSs are a tool, and by themselves do not contain any pedagogical components. The development of sound pedagogical strategies for using CCSs has been the focus of the Physics Education Research Group (PERG) at UMass–Amherst since 1993. UMass PER researchers have published articles on effective uses of CCSs in teaching introductory science (see below). In addition, PER members have conducted numerous workshops with UMass faculty to help them make the transition from student-passive, lecture-style instruction, to student-active, CCS-based instruction. Currently PER continues to provide ongoing technical and pedagogical support to instructors using CCSs in the Physics and Biology departments.

Thus far, 10 introductory courses across four departments at UMass (two courses in Sociology, one in economics, two in biology, and six in physics) have used CCS's to teach large introductory courses. In all cases, both instructors and students have had a very positive experience.

References

R.J Dufresne, W.J. Gerace, W.J. Leonard, J.P. Mestre, and L. Wenk, "Classtalk: A classroom communication system for active learning" *J. of Computing in Higher Educ.* **7,** 3–47 (1996).

J.P. Mestre, W.J. Gerace, R.J. Dufresne, and W.J. Leonard, "Promoting active learning in large classes using a classroom communication system," in E.F. Redish and J.S. Rigden, eds., *The Changing Role of Physics Departments in Modern Universities: Proceedings of International Conference on Undergraduate Physics Education* (American Institute of Physics, Woodbury, NY, 1997), pp. 1019–1036.

L. Wenk, R. Dufresne, W. Gerace, W. Leonard, and J. Mestre, "Technology-assisted active learning in large lectures," in C. D'Avanzo and A. McNichols, eds., *Student-active Science: Models of Innovation in College Science Teaching* (Saunders College Publishing, Philadelphia, PA, 1997), pp.431–452.

Website: *http://umperg.physics.umass.edu/*

L. Video Analysis in the Physics Laboratory

Dean Zollman, Kansas State University

Over the past 15 years video has become a common tool for analysis in the physics laboratory. When students collect data from an event recorded on video, they are using real events to help them understand how the motions are visualized. Interactive video aids students in understanding a variety of complex situations by enabling them to manipulate and measure variables. Data collection can be partially automated while nonlinear video provides flexibility in interactivity. Newer uses of video combined with simulation and modeling tools help students create visual but abstract models of physical processes. These methods provide new pedagogical tools for physics students and offer a much broader learning opportunities.

References

Dean Zollman and Robert Fuller, "Teaching and learning physics with interactive video," *Physics Today* **47** (4), 41–47 (1994).

Lawrence Escalada, Dean Zollman, and Robert Grabhorn, "Applications of interactive digital video in a physics classroom," *J. of Educ. Multimedia and Hypermedia* **5**, 73-97 (1996).

Lawrence T. Escalada and Dean Zollman, "An investigation on the effects of using interactive video in a physics classroom on student learning and attitudes," *J. of Res. in Science Teaching* **34**, 476–489 (1996).

Dean Zollman, "Millikan Lecture 1995: Do they just sit there? Reflections on helping students learn physics," *Am. J. Phys.* **64**, 114–119 (1996).

Teresa Larkin-Hein and Dean Zollman, "Digital video, learning styles, and student understanding of kinematics graphs," *J. of SMET Educ.* **1/1**, 4–17 (2000).

Priscilla Laws and Hans Pfister, "Using digital video analysis in introductory mechanics projects," *Phys. Teach.* **36,** 282–287 (1998).

Dean Zollman "Physics" in *Handbook on Information Technologies for Education and Training*, H.H. Adelsberger, B. Collis, and J.M. Pawlowski, eds. (Springer-Verlag, Berlin, 2002), pp 459–470.

Kansas State University Physics Education Research Group: *http://www.phys.ksu.edu/perg/*

Vidshell 2000 (Doyle Davis), *http://webphysics.tec.nh.us/vidshell/clips.html*

VideoPoint, *http://www.lsw.com/videopoint/*

World in Motion, *http://members.aol.com/raacc/wim.html*

DAVID–Digitale Auswertung von Videos (in German)
http://www.physik.uni-muenchen.de/didaktik/Computer/DAVID/david.htm

Interactive Screen Experiments (in English & German), *http://bifrost.physik.tu-berlin.de/ibe/index.html*

M. Introductory Physics at a Large Research University

Gary Gladding, University of Illinois at Urbana–Champaign

The introductory physics courses at the University of Illinois at Urbana–Champaign have been completely revised in the last five years. The thrust of the revision was to integrate all aspects of a course using active-learning methods based on physics education research in a team-teaching environment.

References

The revisions are documented at:
http://www.physics.uiuc.edu/Education/Course_Revision.html.

A paper describing the project can be found at:
http://www.physics.uiuc.edu/Education/Course_Revision.html

N. RealTime Physics

David Sokoloff, University of Oregon and Ron Thornton, Tufts University

RealTime Physics (RTP) is an introductory laboratory curriculum for those desiring a complete, sequenced set of active learning laboratory activities for an entire semester or quarter, without changing the traditional course structure of lectures and labs. RTP is based on physics education research, makes use of a learning cycle of predictions, observations, comparison and conclusions, and focuses on conceptual and quantitative understanding. Microcomputer-based tools are used extensively, and computers are also used for modeling, data analysis, and simulations. The activities are written generically—using Experiment Configuration Files—so that they are not dependent on a particular hardware and software package. Module 1: Mechanics, Module 2: Heat and Thermodynamics and Module 3: Electric Circuits are published by Wiley. Light and Optics is under development.

References

Ronald K. Thornton and David R. Sokoloff, "RealTime Physics: Active Learning Laboratory," in E.F. Redish and J. R. Rigden, eds., *The Changing Role of the Physics Department in Modern Universities, Proceedings of the International Conference on Undergraduate Physics Education* (American Institute of Physics, Woodbury, NY, 1997), pp. 1101–1118.

Ronald K. Thornton and David R. Sokoloff, "Assessing student learning of Newton's laws: The force and motion conceptual evaluation and the evaluation of active learning laboratory and lecture curricula," *Am. J. Phys.* **66,** 338–352 (1998).

http://ase.tufts.edu/csmthttp://wiley.com/college/sokoloff-physics

O. Interactive Lecture Demonstrations

David Sokoloff, University of Oregon and Ron Thornton, Tufts University

Interactive Lecture Demonstrations (ILDs) are designed to enhance conceptual learning in large (and small) lectures. They are also useful in classrooms where only one computer is available. ILDs are based on physics education research, make use of a learning cycle of predictions, observations, comparison and conclusions, focus on conceptual understanding, and most make use of microcomputer-based laboratory (MBL) tools. The ILD procedure involves students recording individual predictions of the outcomes of simple experiments on a Prediction Sheet (which is collected), discussing their predictions with neighbors and then comparing their predictions to the actual results displayed for the class with the MBL tools. Interactive Lecture Demonstrations in Motion, Force and Energy are available from Vernier Software and Technology. ILDs in other areas are under development.

References

David R. Sokoloff and Ronald K. Thornton, "Using interactive lecture demonstrations to create an active learning environment," *Phys. Teach.* **35** (6), 340 (1997).

Ronald K. Thornton and David R. Sokoloff, "Assessing student learning of Newton's laws: The force and motion conceptual evaluation and the evaluation of active learning laboratory and lecture curricula," *Am. J. Phys.* **66,** 338–352 (1998).

Websites: *http://ase.tufts.edu/csmt and http://www.vernier.com/cmat/ild.html*

P. Studio Physics

Karen Cummings, Southern Connecticut State University

"Studio" teaching is a pedagogical approach rather than a specific curriculum. Developed and refined at Rensselaer between 1995 and 2002, the Studio approach integrates lectures, hands-on activities and instruction in problem solving in each class meeting. A premium is placed on student interactions within groups and with research-active professors. Extensive use of technology helps to make this approach effective in producing student learning and manageable for use at research universities.

References

For more information, see the references below or contact Karen Cummings at karen@rpi.edu or Jack M Wilson at JackMWilson@JackMWilson.com .

J. Wilson, "The CUPLE physics studio," *Phys. Teach.* **32** (9), 518–523 (1994).

K. Cummings and J. Marx, "Evaluating innovations in studio physics," *Am. J. Phys.* (*Phys. Ed. Res. Supplement*) **67** (7), S38-S44 (1999).

Q. Other Web Resources

Here we mention a few websites that offer collections of information on undergraduate physics and links to many other undergraduate physics web resources:

- The American Association of Physics Teachers maintains a website "Physical Sciences Resource Center," which contains much information and many links to other sources about undergraduate physics. *http://www.aapt.org*

- Project Galileo at Harvard University contains a collection of resources for undergraduate physics. *http://galileo.harvard.edu/lgm/pi*

- The large-scale digital library project comPADRE for physics is under development as of this writing (early 2003). Preliminary materials are expected to be ready through the AAPT website during the fall of 2003.

Appendix II.
Undergraduate Physics Reading List

(compiled by J. D. Garcia)

NTFUP's goals include encouraging awareness of the changing educational environment, promoting best practices in undergraduate physics education and providing mechanisms for greater dialog among physicists concerning undergraduate physics education. As a means of encouraging discussion and as a starting point for thinking about what has worked at various places, we have assembled an admittedly incomplete and selective set of articles and materials from the literature dealing with the teaching of physics and with physics education research. The resource letter on physics education research [McDermott and Redish, 1999] is a much larger bibliography on literature on the subject.

We encourage you to read this material and discuss it with your colleagues. These readings are intended to be only a starting point for discussions. Indeed, were we to include material on all programs, practices, and innovations that we deem worthwhile, the list would be prohibitively long.

Patricia Heller, Ronald Keith, and Scott Anderson, "Teaching problem solving through cooperative grouping. Part 1: Group vs. individual problem solving," *Am. J. Phys.* **60,** 627 (1992).

Patricia Heller, Ronald Keith, and Scott Anderson, "Teaching problem solving through cooperative grouping. Part 2: Designing problems and structuring groups," *Am. J. Phys.* **60,** 637 (1992).

William J. Leonard, Robert J. Dufresne, and Jose P. Mestre, "Using qualitative problem-solving strategies to highlight the role of conceptual knowledge in solving problems," *Am J. Phys.* **64,** 1495 (1996).

Robert C. Hilborn, "Guest Comment: Revitalizing undergraduate physics—Who needs it?" *Am. J. Phys.* **65,** 175 (1997).

Edward F. Redish, "Millikan Lecture 1998: Building a science of teaching physics," *Am. J. Phys.* **67,** 562 (1997).

Eric Mazur, *Peer Instruction,* Chapter 2: Concepttests, (Prentice Hall, Upper Saddle River, NJ, 1997).

Lillian McDermott, "Bridging the Gap Between Teaching and Learning: the Role of Research," in *Proceedings of the International Conference on Undergraduate Physics Education,* CP399, edited by E.F. Redish and J.S. Rigden, (AIP Press, Woodbury, NY, 1997), pp. 139–165.

Frederick Reif, "How Can We Help Students Acquire Effectively Usable Physics Knowledge?" in *Proceedings of the International Conference on Undergraduate Physics Education*, CP399, edited by E.F. Redish and J.S. Rigden, (AIP Press, Woodbury, NY, 1997), pp. 179–195.

Rosanne Di Stefano, "Where an Instructor's Dreams Meet Reality: Total Available Student Time," in *Proceedings of the International Conference on Undergraduate Physics Education, CP399,* edited by E.F. Redish and J.S. Rigden, (AIP Press, Woodbury, NY, 1997), pp. 225–239.

Ronald K. Thornton, "Conceptual Dynamics: Following Changing Student Views of Force and Motion," in *Proceedings of the International Conference on Undergraduate Physics Education,* CP399, edited by E.F. Redish and J.S. Rigden, (AIP Press, Woodbury, NY, 1997), pp. 241–265.

Richard R. Hake, "Evaluating Conceptual Gains in Mechanics: A Six Thousand Student Survey of Test Data," in *Proceedings of the International Conference on Undergraduate Physics Education,* CP399, edited by E.F. Redish and J.S. Rigden, (AIP Press, Woodbury, NY, 1997), pp. 595–603.

Fred Goldberg, "How Can Computer Technology be Used to Promote Learning and Communication Among Physics Teachers?" in *Proceedings of the International Conference on Undergraduate Physics Education,* CP399, edited by E.F. Redish and J.S. Rigden, (AIP Press, Woodbury, NY, 1997), pp. 2375–2392.

Richard Hake, "Interactive-engagement versus traditional methods: A six-thousand-student survey of mechanics test data for introductory physics courses," *Am. J. Phys.* **66**, 64–74 (1998).

Lillian C. McDermott and Edward F. Redish, "Resource Letter: PER 1: Physics education research," *Am. J. Phys.* **67,** 755–767 (1999).

Ruth H. Howes and Robert C. Hilborn, "Guest Comment: Winds of change," *Am. J. Phys.* **68,** 40 (2000).

Corinne A. Manogue, Philip J. Siemens, Janet Tate, Kerry Browne, Margaret Niess, and Adam J. Wolfer, "Paradigms in physics: A new upper-division curriculum," *Am. J. Phys.* **69,** 978 (2001).

Edward F. Redish, *Teaching Physics* (John Wiley & Sons, New York, 2003).

Appendix III.
Presentations and Articles on SPIN-UP

A. Presentations

1. APS Meeting, Long Beach, CA, April 2000.

What's Happening in Undergraduate Physics Revitalization?
Robert C. Hilborn (University of Nebraska-Lincoln)

The American Association of Physics Teachers, the American Physical Society, and the American Institute of Physics have recently launched the National Task Force on Undergraduate Physics. The Task Force's initial activities are also supported by a planning grant from the Exxon Education Foundation. The goal of the Task Force is to coordinate a number of activities led by AAPT, APS, AIP and others to foster the "revitalization" of undergraduate physics programs across the country. The Task Force will also provide advice about new activities aimed at undergraduate physics. This effort emphasizes all aspects of undergraduate physics including the recruitment and mentoring of students, providing strong courses for physics majors, other science majors, nonscience majors and pre-service K–12 teachers, engaging students in research, and preparing students for a diverse set of careers. The Task Force focuses on the department as the fundamental unit for undergraduate education change while recognizing that innovations must be adapted to suit local needs. In this talk I will give some background of the events leading up to the establishment of the Task Force. I will also discuss some of the activities aimed at revitalizing undergraduate physics and plans for future programs under discussion by the Task Force.

2. AAPT Meeting, Guelph, Ontario August 1, 2000.

What's Happening in Undergraduate Physics Revitalization?
Robert C. Hilborn (Amherst College)

AAPT, the American Physical Society (APS), and the American Institute of Physics (AIP) have recently launched the National Task Force on Undergraduate Physics. The Task Force's initial activities are also supported by a planning grant from the ExxonMobil Foundation. The goal of the Task Force is to coordinate a number of activities led by AAPT, APS, AIP, and others to foster the "revitalization" of undergraduate physics programs across the country. The Task Force will also provide advice about new activities aimed at undergraduate physics. This effort emphasizes all aspects of undergraduate physics including: recruiting and mentoring students; providing strong courses for physics majors, other science majors, nonscience majors, and pre-service K–12 teachers; engaging students in research; and preparing students for a diverse set of careers. The Task Force focuses on the department as the fundamental unit for undergraduate education change while recognizing that innovations must be adapted to suit local needs. In this talk I will give some background of the events leading up to the establishment of the Task Force. I will also discuss some of the activities aimed at revitalizing undergraduate physics and plans for future programs under discussion by the Task Force.

3. APS Meeting, Washington, DC, April, 2001.

Building Undergraduate Physics Programs for the 21st Century
Robert C. Hilborn (Amherst College)

Undergraduate physics programs in the United States are under stress because of changes in the scientific and educational environment in which they operate. The number of undergraduate physics majors is declining nationwide; there is some evidence that the "best" undergraduate students are choosing majors other than physics, and funding agencies seem to be emphasizing K–12 education. How can physics departments respond creatively and constructively to these changes? After describing some of the details of the current environment, I will discuss the activities of the National Task Force on Undergraduate Physics, supported by the American Institute of Physics, the America Physical Society, the American Association of Physics Teachers and the ExxonMobil Foundation. I will also present some analysis of Task Force site visits to departments that have thriving undergraduate physics programs, pointing out the key features that seem to be necessary for success. Among these features are department-wide recruitment and retention efforts that are the theme of this session.

4. PKAL Summer Institute, Williamsburg, VA, June 2–5, 2002.

Brief presentation to all participants by Robert C. Hilborn (Amherst College).

5. AAPT/APS Physics Department Chairs Meeting, College Park, MD, June, 2002.

Report on Site Visits to Physics Departments
Ruth H. Howes (Ball State University)

6. NSF Physics, MPS, and DUE program officers, Arlington, VA, June 13, 2002.

Summary of SPIN-UP results presented by Robert C. Hilborn (Amherst College).

7. AAPT 2002 Summer Meeting, Boise, ID, August 5, 2002.

The National Task Force on Undergraduate Physics and SPIN-UP
Robert C. Hilborn (Amherst College)

The National Task Force on Undergraduate Physics, a joint effort of the American Association of Physics Teachers, the American Physical Society, the American Institute of Physics, and Project Kaleidoscope, was established in 1999 to provide advice to the physics professional societies and the physics community at large about the state of undergraduate physics. After reviewing some of the background leading up to the establishment of the Task Force, I will describe the project Strategic Programs for Innovations in Undergraduate Physics (SPIN-UP), a Task Force effort funded by the ExxonMobil Foundation. SPIN-UP focuses on site visits to about 20 colleges and universities that have thriving undergraduate physics programs and a survey, conducted in cooperation with AIP, of all undergraduate physics departments in the country. I will discuss the common features, identified from the site visits, found in departments that have thriving undergraduate physics programs.

The SPIN-UP Survey of Undergraduate Physics Programs
Kenneth Krane (Oregon State University)

In spring 2002, SPIN-UP (Strategic Programs for Innovations in Undergraduate Physics) of the National Task Force on Undergraduate Physics conducted (through the American Institute of Physics) a survey of undergraduate physics programs throughout the United States. Among the information that the survey form was designed to elicit were: (1) undergraduate curricula, including the character of the department's "standard" degree track and any alternative degree tracks that are available; (2) activities for recruiting undergraduate majors; (3) interactions between faculty and physics majors, including advising and mentoring as well as informal contacts; (4) alumni relations; and (5) curricular reform efforts. In addition to gathering information, the survey asked departments to evaluate the success of these activities and to discuss the current strengths and needs of the department. We will review the survey document and present the results analyzed to date.

Using the Results: Next Steps and Getting Involved
Ruth Howes (Ball State University)

SPIN-UP (Strategic Programs for Innovations in Undergraduate Physics) has studied the condition of undergraduate physics programs in all kinds of colleges and universities through site visits and a survey, the results of which have been presented in this session. We have focused on thriving departments with successful undergraduate programs. Not all undergraduate physics programs are thriving. The National Task Force on Undergraduate Physics is preparing to use the results of SPIN-UP to help other departments change constructively. We report on future plans and opportunities for AAPT members to become involved in improving undergraduate physics programs.

8. European Union Physics Departments Meeting, Varna, Bulgaria, September 7, 2002.

Presentation by Ruth H. Howes (Ball State University)

9. NSF-Corporate Foundation Alliance Meeting, Arlington, VA, October 23, 2002.

Presentation by Robert C. Hilborn (Amherst College).

10. Mid-West Physics Department Chairs Meeting, Chicago, November 3, 2002.

SPIN-UP Results and Analysis
Robert C. Hilborn (Amherst College)

B. Articles about the Task Force, and SPIN-UP and Related Activities.

"Revitalizing physics education," *Physics Today* **53** (4), 59–60 (April 2000). Brief notice about the formation of the Task Force.

"APS, AIP, and AAPT launch task force on undergrad physics," *APS News* **9** (4) (April 2000).

"The physics department 'Cosmo Quiz'," *APS News* **9** (4) (April 2000).

D. E. Neuenschwander, "What does 'Physics Revitalization" mean?" *Reveille* 2000.

Ruth H. Howes and Robert C. Hilborn, "Winds of change," *Am. J. Phys.* **68,** 401–402 (2000).

Robert C. Hilborn, "The National Task Force on Undergraduate Physics: Some FAQs," *APS Forum on Education Newsletter* (Spring/Summer 2000).

Carl Wieman, "A Modest proposal: Recruit undergraduate majors," *APS News* **10** (5) (May 2001).

Robert C. Hilborn, "The National Task Force on Undergraduate Physics," National Research Council Board on Physics and Astronomy *BPA News* (June 2001).

"Amherst Professor Hilborn to head National Task Force on Undergraduate Physics," *Amherst College Notes*, August 30, 2001.

"SPIN-UP seeks undergraduate programs to host site visits," *APS News* **12** (12) (December 2001).

Ken Krane, "What produces a thriving undergraduate physics program?" *APS News* 11 (11) (November, 2002).

Appendix IV.
Site Visit Volunteers

Mary Alberg
Seattle University

Teresa Burns
Coastal Carolina University

Ruth W. Chabay
*Carnegie Mellon University
(now at North Carolina State
University)*

Cliff Chancey
University of Northern Iowa

Wolfgang Christian
Davidson College

Robert Ehrlich
George Mason University

William Evenson
Brigham Young University

Andrew Gavrin
*Indiana University-Purdue
University*

Gary Gladding
University of Illinois

Kenneth Heller
University of Minnesota

Dennis Henry
Gustavus Adolphus College

Theodore Hodapp
Hamline University

Donald Holcomb
Cornell University

William Ingham
James Madison University

Patrick Kenealy
*Cal State University Long
Beach*

Randall D. Knight
*Cal Poly State University, San
Luis Obispo*

John Knox
Idaho State University

Jean Krisch
University of Michigan

Priscilla Laws
Dickinson College

Ramon Lopez
University of Texas El Paso

Catherine Mader
Hope College

Mary Beth Monroe
*Southwest Texas Junior
College*

Kathie Newman
University of Notre Dame

Thomas Olsen
Lewis and Clark College

Richard Peterson
Bethel College

Rick Robinett
Pennsylvania State University

Lyle Roelofs
Haverford College

Warren Rogers
Westmont College

Richard Saenz
Cal Poly State University

James R. Sites
Colorado State University

David Sokoloff
University of Oregon

Patricia Sparks
Harvey Mudd College

Paul Stanley
*California Lutheran
University*

Conley Stutz
Bradley University

Doyle Temple
Hampton University

Michael Thoennessen
Michigan State University

Ed Thomas
*Georgia Institute of
Technology*

Jan Tobochnick
Kalamazoo College

Dean Zollman
Kansas State University

Appendix V.
Site Visit Documentation

This appendix contains the documents that were used in setting up and running the site visits.

A. Definition of a Thriving Undergraduate Physics Program

1. The number of majors is stable at a level that the department and the administration consider satisfactory or shows significant and sustained growth toward that number.

2. Morale is high for both faculty and students. They are engaged with physics, and the atmosphere within the department is collegial. Faculty regularly evaluate and respond to the changing needs of their students both majors and students in service courses, and work to enhance their skills as teachers. They seek to improve the experiences they offer their undergraduate students by constantly updating the departments' curriculum and by involving undergraduate colleagues in research.

3. Graduates find good jobs or obtain admission to graduate programs both in physics and in other fields. The department actively supports the professional development of its students by activities such as making information available about diverse careers, arranging for internships, or working with industries in an industrial advisory committee.

4. The college or university in which the department is situated respects the department, and all students find its programs attractive. Here "all" includes students enrolled in service and general education courses as well as physics majors and minors.

5. The department faculty work as a team to provide excellent undergraduate education. The majority of the faculty consider undergraduate teaching very important and honor their colleagues who do it even if they personally are not actively involved. The department invests resources not only in major courses but also in service and general education courses.

6. The department regards both undergraduate students and staff members as important members of the departmental team. Their voices are heard in making departmental decisions.

7. The department attracts and retains minorities and women as physics majors.

8. The department recognizes its responsibility to promote excellence in physics education for all K–12 students. This responsibility may be expressed through a variety of activities, for example: direct education of pre-service teachers; supportive involvement in physics or physical science courses or curricula for pre-service teachers (whether or not these courses are not taught by the physics department); in-service programs for local teachers; or outreach activities for local teachers and students.

B. Letter to the Site Visit Chair

Dear (Department Chair):

The purpose of the site visits of the National Task Force on Undergraduate Physics is to learn what makes a thriving undergraduate physics program. We are particularly interested in innovations in physics departments that could be widely duplicated. The site visits are supported by Strategic Programs for Innovations in Undergraduate Physics (SPIN-UP) through funding from the ExxonMobil Foundation.

The Task Force seeks to answer two questions:

1. What are the activities that position physics departments for **success** in producing more majors, placing graduates in a variety of interesting careers, and playing a productive and significant role in the academic life of the institution through both service courses, general studies courses, outreach programs, and so on?

2. What are the essential conditions within a department needed to promote a constructive and creative response to environmental change, and what events or pressures combined to stimulate those responses?

Some secondary (but important) issues:

Some departments have a large fraction of their faculty involved in innovating their undergraduate programs. However, many departments have only a small fraction of the faculty involved actively in undergraduate programs beyond routine meeting of classes. In this case, it is critical that other members of the department support them in such tangible ways as promotion and tenure. What is the minimum number of active faculty needed for excellence in an undergraduate program, particularly for making substantive changes? What support from the rest of the department is absolutely essential? How long does it take to produce lasting change within a physics department? How do the department and the institution measure success, particularly the effect of innovations, in the undergraduate program? How are the resources and faculty time needed to create a thriving undergraduate program balanced against other demands on the department and the faculty?

The documentation submitted by the department before the visit should provide data on what the department thinks it has accomplished. The site visit is needed to look for elements such as morale of faculty and students and institutional support that do not appear in formal reports. The visit is not intended to evaluate directly the strengths and weaknesses of the department's program. We do, however, want to achieve a realistic picture of what was done, how it was done, and how it is working. The eventual goal is to be able to characterize those elements that are important (or in some cases crucial) for planning, developing, implementing, and sustaining successful undergraduate physics programs. We must keep in mind that what constitutes a "thriving" program is subject to local interpretation though, of course, there will be many features common to all physics programs.

The attached contract explicitly states the terms under which the site visit will be conducted. Please sign it, return it to me, and keep a copy for your files. Also attached are several questions whose answers should be provided to the Task Force before the site visit. The Site Visit Team will consist of three physicists including one member of the Task Force. We will try to select members of the team from institutions geographically close to yours. Should you desire it, a member of the Site Visit Team will present a colloquium to your department. More information

about the Task Force and its membership can be found on the AAPT website:
http://www.aapt.org/Projects/ntfup.cfm.

The Task Force appreciates your agreeing to participate in the SPIN-UP site visit program. Your contribution will help other physics departments design constructive responses to the changing environments in which they find themselves.

C. Contract

The Task Force makes the following agreement with the Physics Department at _____:

The department will cover all local expenses (housing, meals and local transportation during the visit) for the three-member site visit team.

The department will make appropriate hotel reservations for the site visit team. SPIN-UP will cover all travel expenses for the site visit team.

The department will provide the site visit team with written responses to a set of questions about the department's undergraduate program at least two weeks prior to the site visit.

In consultation with the site visit team leader, the department chair will set up a schedule of appointments with small groups of faculty (both in the department and outside the department as appropriate), students (both majors and nonmajors and special groups such as engineers, pre-service teachers, etc.), clerical and technical staff, and administrators.

After the site visit, the site visit team will provide the department chair with a written report of the team's findings within three weeks of the site visit. The chair will have one week to correct factual errors in the draft and return it to the chair of the site visit team. The report in final form will be submitted to the department chair and NTFUP. The report is written for the department chair. The chair may share the report with other members of the department and with the institution's administration at the discretion of the chair. The Task Force will seek the permission of the chair before using any of the data in the report in a way that links the data directly to the department. The Task Force may ask for additional data and comments as it prepares a Case Studies document.

D. Departmental Questionnaire for Task Force Site Visits

The Task Force site visit will be much more productive both for the Task Force and for the department if the site visit team members have some information about the department in advance of the actual visit. Please provide the information described below. (If you have this information in a different format, for example, for a recent departmental review or self-study, please feel free to substitute that report for the format given below.) We emphasize that this visit is not a usual departmental review. We are interested in the steps you have taken to ensure that your department's undergraduate program is truly thriving. The site visit team wishes to collect data ahead of time and spend site visit time talking with faculty, staff, students and administrators.

1. Personnel information

Please list:

    A.   Faculty by rank and give years in service and a brief statement of research areas

    B.   The number of support staff, (for example, departmental assistants, lecture demo support staff, lab instructors, technical staff, machine shop staff, electronics shop staff, etc.) and indicate if these staff are full-time or part-time.

2. Information about students

Please list:

    A.   The number of majors you have by class (first-year, sophomores, juniors, seniors) and the number of majors you have graduated each year for the last 10 years. Any data you have on entering physics majors or enrollment by class for different years would be helpful, as would information on demographic characteristics of your students. (For example, do you have a large number of nontraditional or transfer students? How many minority or women students are physics majors? Do your students come from public schools? Rural schools? Private schools? Do you have any information on their SAT scores or their high school grade point averages? Have most of them taken AP physics?) You don't need to undertake major research for this questionnaire, but data on the characteristics will allow the SVT to acquire a clearer picture of your department.

    B.   The typical enrollments in each of the undergraduate courses offered by your department (precise figures are not necessary). It would be helpful to have a brief phrase describing each course and its primary audience. Alert us to any historical trends in the data.

    C.   Typical career paths for your physics majors. Roughly what fraction go directly into the workforce, to K–12 teaching, to graduate school, to professional school, etc. Again alert us to any historical trends in those data.

    D.   Research participation and TA opportunities for undergraduates in your department.

    E.   Other opportunities for physics majors outside the classroom (e.g. an active SPS chapter, a student lounge, tutoring, etc.)

3. Provide a brief narrative about your undergraduate physics program (including the program for majors and courses for nonmajors), particularly focusing on what you consider to be the most important components and novel features that you believe are particularly successful. We would also like to learn about how the department planned for and implemented innovative features and how they are being evaluated and sustained. The following questions should be addressed:

    A.   What changes have you made during the last five years to improve the experiences of undergraduate students in your department's program?

B. How did you make these changes? Specifically: Why did you embark on change?

C. How did you recruit faculty to work on the new programs?

D. How did you obtain resources to support change? What added resources were you able to obtain?

E. What evidence do you have that your department is thriving and that your activities produce success?

4. If you have other general information about your undergraduate program including recruiting brochures, course catalog information, course or faculty evaluation forms, and so on, we would appreciate receiving copies of that information.

5. What academic or psychological services (such as tutoring or help with test anxiety) does your department or your university provide to students? What services does your university provide to students that are particularly useful to your department? For example, some physics departments benefit greatly from the active recruiting programs of their colleges.

6. Does your department play a significant role in the preparation of K–12 teachers? Describe that role and the department's interactions with the school of education (or the appropriate group within your institution).

7. Please feel free to send along other information that you believe might give us a good picture of your department and its program.

E. Site Visit Protocol

Before the Visit:

The department chair and project directors agree on terms of visit. The chair returns signed contract to the project director who is coordinating the visit.

1. The project directors select a site visiting team (SVT) and coordinate dates with the department being visited. The SVT will consist of three people, one of whom will be a member of the Task Force. The project directors will inform the chair of the membership of the SVT, coordinate travel for the visitors, and assure that the department has made appropriate hotel and meal arrangements. The project directors will check with site visitors regarding special dietary and room arrangements. Two weeks before the visit, the project directors send each member of the SVT a packet of information about SPIN-UP.

2. The department prepares answers to a questionnaire given to the chair along with the contract and sends it to the project directors at least two weeks before the scheduled visit. The project directors distribute copies of the questionnaire to members of the SVT. The department will also supply copies of catalog materials and, at its discretion, copies of recent external reviews or self-study evaluations.

3. A week before the visit, the department chair sends the project directors a schedule of activities planned for the site visit that the project director then sends to the SVT.

4. Two or three days before the visit, the SVT should schedule a conference, call, or exchange email to identify issues and questions about the department.

5. If a colloquium by one of the members of the SVT is part of the visit, the project directors will ensure that the department receives a title in a timely manner.

It is essential that the site visiting team meet as a group before starting the visit. If possible, the team should plan to arrive in late afternoon and meet in the evening. Otherwise, a breakfast meeting should be scheduled. The department chair may be invited to attend all or part of this meeting. The purpose of this meeting is to discuss the written material and prepare a strategy for finding answers to questions that arise.

During the visit (which is expected to last about one and a half days), the SVT should meet with: the chair, the coordinator of undergraduate programs, faculty in circumstances where informal discussion is possible, at least two groups of undergraduate students (both majors and nonmajors), the dean or other appropriate administrator, key departmental staff, and others selected by the department. The number of formal presentations to the team should be kept to a minimum with all necessary factual information presented in the materials sent out before the visit.

At the end of the visit, the SVT should meet in executive session to discuss their report. If it seems appropriate, the SVT may meet with the chair to summarize their findings.

F. After the Visit

The chair of the site visit team appointed by the project directors is primarily responsible for drafting the team's report to the department chair and NTFUP. The initial draft should be circulated electronically to the SVT for comment and correction and then (within three weeks of the visit) sent to the department chair for correction of factual errors. The department chair will then have one week to respond with reports of errors or omissions. The report should be in final form and submitted electronically to the SPIN-UP project directors and the department chair not more than one month after the visit. The project directors will share the report with the members of NTFUP who are asked to keep it confidential.

G. Sample Schedule

National Task Force on Undergraduate Physics
Visit to Colorado School of Mines Physics Department
October 5–6, 2000

Thursday Oct. 5

8:30: Meeting with Dept. Head (Prof. McNeil, Room 325)

Greetings/Orientation

9:00: Meeting with Pres. Trefny — Institutional perspective

10:00: Tour of department and review of data (McNeil)

11:00: Meeting with Freshman/Sophomore Physics majors (Room 347)

12:00: Lunch   (Table Mountain Inn with Physics faculty)

1:30–2:30: Meeting with Junior/Senior Physics Majors (Room 347)

2:30–3:30: Meeting with half of physics faculty (Room 335)

3:30–4:30: Meeting with other half (Room 335)

7:00: Dinner (240 Union with Physics faculty)

Friday Oct. 6

8:30: Task Force team meeting (Room 335 will be available, if needed)

9:00: Meet with D. Williamson (previous Head)

10:00: Exit interview with Dept. Head (Room 325)

Appendix VI.
Formative Evaluator's Report

Developing a Framework for Creating Thriving Physics Departments:
A Report to the National Task Force on Undergraduate Physics
by
Charles R. Payne

Introduction

Over the past 10 years, the number of physics majors in university physics departments has been steadily declining. The reasons for the decline are readily apparent. According to Krane, Department of Physics, Oregon State University, the decline in undergraduate physics enrollments in the 1990s has been well documented.

> The number of baccalaureate physics degrees awarded per year in the United States dropped by about 25%, from about 5000 per year in 1990 to about 3800 per year in 1999. Simultaneously, the total number of bachelor's degrees was *increasing*, from about 1,000,000 per year to 1,200,000. During this period, the fraction of physics degrees awarded thus fell from 0.5% of total bachelor's degrees to 0.3%. Although there is perhaps evidence of a small uptick in the data for the past year or two, it is not clear that this represented a trend and even less clear that it can be sustained to reverse the decline of the past decade.

To respond to the issue of the decreasing number of physics graduates, an 11-member National Task Force on Undergraduate Physics (NTFUP) was formed in 1999 under the joint sponsorship of the American Physical Society, the American Association of Physics Teachers, and the American Institute of Physics. NTFUP is a group of physicists who are befitting of the term a "community of learners." They are well organized in that there appears to be representation from all of the major physics organizations. Among its members are two Noble Prize winners, physics department chairpersons from leading institutions, and other well-known physicists. As one would surmise, NTFUP is a very politically astute group.

The NTFUP studied how the environment has changed for undergraduate physics programs. They also investigated the constructive and innovative responses that departments have taken toward the changing environment. NTFUP operated on the assumption that the primary cause of the decline in the number of physics majors is due to the changes within the environment. They also believe that if physics departments are to thrive, then the physics community must respond to the changes. In his essay, "What Produces a Thriving Undergraduate Physics Program," Krane discussed the issues related to the decline in enrollment of undergraduate physics programs.

Although physics graduates are declining nationwide, a few physics departments can be described as thriving. At the request of the department chairs of 21 thriving departments, visitation teams consisting of three physicists were formed to visit these departments. The visits lasted on the average of one and a half days. Each team wrote a summary report of their visit. A framework for producing a thriving department was developed inductively from observations of common themes found in the reports.

A. Credibility of the Framework

The framework of a thriving department was arrived at inductively by extracting data from the reports of the 21 visitation teams. The framework is described as general characteristics, similar trends, and common themes of thriving physics departments. Although some team members visited more than one department, a total of 54 different physicists were involved in the visitations. The large number of highly trained physicists who made similar observations of thriving schools validated the reported observations and strengthened the reliability and validity of the framework.

The schools and the environments within which these thriving departments existed were very diverse. With respect to student enrollment and the number of departmental faculty members, the colleges and universities ranged from some of the smallest to some of the largest in the country. While all of the schools were recognized as quality institutions, it is salient to mention that some of these institutions were recognized as being among the country's most selective schools. The participating schools included private, public, and religious institutions as well as large urban and rural campuses. Geographical diversity was also represented in that most sections (East, Midwest, West, Northwest, and Southwest) were represented. The Southeast was the only geographical area that was not represented. These states included Alabama, Florida, Georgia, South Carolina, Tennessee, Mississippi, and Louisiana. It is also important to reiterate that the departments that participated in the visitations were identified as being successful in increasing the number of physics majors over the past few years.

Regardless of differences between the participating schools, essentially the same practices were being implemented in all of the thriving physics departments. From observations made by reading the reports, it has been concluded that there were certain practices common to all thriving physics departments. These practices are discussed below as key elements in the framework that other departments might be able to follow when attempting to change the departmental environment to be more inviting to students.

B. Elements of the Framework for Creating a Thriving Physics Department

A total of fourteen (14) elements were derived from the review of the reports. For the purposes of this report, each of elements will be listed and discussed in turn. The evidence that supports the element will also be provided along with reviewer comments where appropriate.

Element 1: Thriving physics departments have a reputation as being first rate in the types of academic programs that are offered, the pedagogical skills of the faculty, and the nurturing environment established by the faculty.

Evidence: The departments included in the visitations were chosen because they were going against the national trend of having a decline in the number of physics majors. All reports began with a description of the high academic quality of the undergraduate physics programs.

Comments: While it should be a given, it frequently has to be stated that "quality counts" when students choose academic programs. It is often the case when an academic department wants to increase the number of students involved in its program that the immediate fear of the faculty and the public is that the standards will be compromised. Nothing could be farther from the truth with regard to programs in these reports. The students who are attracted to the physics major are high caliber and very capable of determining whether or not they are receiving a high-quality education. They are also students who want to be challenged and they recognize when they are being challenged.

Of the 21 reports, there was only one instance where some members of the physics faculty intimated that quality was being sacrificed for the sake of increasing the number of majors. In this instance some of the faculty had negatively nicknamed one of the new courses "Light Physics."

Element 2: Thriving physics departments offer students both research opportunities and personal involvement with professors.

Evidence: All 21 physics departments indicated that they had research opportunities available for undergraduate students. While research conducted by undergraduates with the guidance of professors was strongly encouraged by all of the departments, approximately half of the departments made it a requirement. There was also a trend for departments that had not focused on undergraduate research opportunities in the past to become more active in making research opportunities for undergraduates more available.

Comments: Research opportunities were being made available in a variety of ways. For example, some of the smaller departments were encouraging their students to gain greater exposure by applying for Research Experiences for Undergraduates (REUs) at larger institutions during the summer. Other departments were beginning to fund their own research projects. All departments including those with the smallest number of faculty members offered some type of research opportunities and experiences for their undergraduate students. Quite often a senior research project was a capstone experience for undergraduate physics majors.

Element 3: Thriving physics departments have faculties whose reputations for having excellent pedagogical skills rank highly for attracting students into the major.

Evidence: It was frequently reported by visiting teams that students regularly commented on the pedagogical skills of the physics faculty. Students' comments were generally unsolicited and compared the physics faculty with faculty members in other departments. Some departments recognized a need for a change in pedagogy through a review of physics education research (PER). Departments also reported that they began to experience growth in the number of majors when the faculty began incorporating research findings from PER into classroom instruction. One of the schools developed an elaborate system to reward excellence in instruction for both faculty and graduate teaching assistants.

Comments: It has been a long-standing belief within the educational community that learning is determined by how people are taught, and that the quality of learning for any student is directly related to the quality of instruction. Excellent instruction, therefore, must be the premise for all changes. In addition, good teaching should be recognized and rewarded.

Element 4: Thriving physics departments have professors who serve, either formally or informally, as advisors.

Evidence: It was stated directly in 10 of the 21 departmental reports that physics professors served as academic advisors. Additionally, another 10 departments implied that the physics professors served informally as academic advisors. One department employed a person with a science degree to serve as an advisor, while only two departments made no mention of an advising component for physics majors.

Element 5: Thriving physics departments have goals that are clearly stated, are well known, and understood by the faculty and staff. The departmental goals are also consistent with the goals of each respective university.

Evidence: It was either directly or indirectly stated in each of the reports that the physics departments had goals that were aligned with the universities' goals. These departmental goals were clearly stated and understood by most faculty members. Visiting teams frequently met with administrators who expressed gratitude for the efforts of the physics department faculty in corroborating the goals of the university.

Element 6: Thriving physics departments actively recruit physics majors.

Evidence: Eighteen of the 21 departments reported some type of direct recruitment of physics majors. Seven of those departments reported having direct contact with and active recruitment in high schools. Three of the most successful physics programs, however, seemed to have a limited involvement with recruiting but were successful in attracting physics majors based on academic reputation alone. However, these same programs reported fostering a strong sense of community for physics majors.

Element 7: Thriving physics departments have departmentalized the practices that have been implemented to attract students. All departmental faculty members reported embracing the efforts that were put forth by a few of its members as valuable to the entire department.

Evidence: This was especially evident from the reports of three of the most highly successful programs. The problem with the lack of departmentalization was made very clear from one example of a department that described an aggressive and successful recruitment program; however, recruitment was done only by one individual. This was particularly problematic last year when this person was on leave and only half of the usual number of students chose physics as a major.

Element 8: Thriving physics departments foster environments where personal involvement of the faculty with individual students is the rule.

Evidence: Faculty involvement with students and their availability to students were elements that were observed in 100% of the departments. According to the reports, students readily responded to these faculties and expressed appreciation for their attitudes. Visiting teams, when describing faculty and student interactions, frequently used the expressions "sense of community" and "a community of learners." As reported by four of the visiting teams, the apparent positive effect that this type of interaction had on students prompted a comment such as the following: "Why can't other departments foster such an atmosphere?"

Comments: Involvement of the faculty was carried out in a number of ways.

An informal atmosphere where faculty are friendly with each other and accepting of students as members of the physics family.

All majors have easy access to faculty and departmental administration.

Majors are made to feel like they are part of the department and the physics endeavor.

Element 9: Thriving physics departments have flexibility in the physics curriculum.

Evidence: Flexibility within the physics curriculum was found to be the rule rather than the exception. Although there are a few departments that continue to hold on to the traditional physics curriculum, the trend is toward more flexible programs. For example, a number of departments have begun to offer a Bachelor of Science degree with fewer courses than the typical Bachelor of Arts.

Comment: There was great flexibility offered in the physics curriculum, but the most common pattern was a physics major combined with one of the other sciences; however, there were some unique combinations. It also appeared as though many departments had begun to reduce the physics requirements in order to accommodate a more flexible physics major. Of course this raised the issue of the standards of courses and programs being compromised. If a program meets the professional needs of a student by offering fewer courses, then it does not inherently lower the quality of the physics program.

Element 10: Thriving physics departments have strong institutional support both financially and academically.

Evidence: According to the departments, institutional support is provided for the physics departments in a number of ways. Some examples include the following: (a) funding undergraduate research projects; (b) supporting department chairs who made crucial decisions about implementing change; (c) granting permission to hire additional faculty; (d) supporting a more flexible physics curriculum; and (e) giving general praise and appreciation for faculty of the physics departments.

Comment: When an institution depends on its faculty to make major changes without financial assistance, it runs the risk of burning out the faculty. The possibility of faculty being over worked was raised in several of the reports, particularly in smaller departments.

Element 11: Thriving physics departments have a chapter of the Society of Physics Students (SPS) and/or other similar organizations.

Evidence: Seventeen departments indicated that they had SPS chapters. Although the activity level of the chapters varied, many of the departments gave high marks to the SPS chapters for creating a nurturing environment. The chapters were also responsible for many activities such as seminars, field trips, recruiting and other social functions. Six of the departments did not have SPS chapters; however, activities similar to SPS were carried out informally.

Element 12: Thriving physics departments are committed to undergraduate physics.

Evidence: All 21 departments indicated that they have made a commitment to undergraduate physics. For some departments, making this commitment was a major step toward creating an environment that would attract more majors. Several of the departments indicated that faculty made a conscious decision not to develop graduate programs but to focus on undergraduate physics. A common theme throughout the reports with regards to undergraduate physics was as follows:

> We learned that the entire faculty is involved in discussions of the undergraduate program at the department's faculty meetings, though revisions of particular courses or the development of new courses is often done by individual faculty members or a small group of faculty.

Element 13: Thriving physics departments have faculty members who are accepting and nurturing of students.

Evidence: The idea of a nurturing environment was a major thread that seemed to help establish bonds between students and faculty. Although departments used differing terminology, each addressed creating a nurturing environment. A representative categorization of one department is as follows:

a. Undergraduate majors get keys to the building and have access to a library, a computer area, and an informal "penthouse" area with a refrigerator, microwave, etc.

b. The physics building layout has faculty offices and labs in close proximity to classrooms and instructional labs. This fortuitous geographical layout encourages students to take advantage of the "open door" policy of most of the department's faculty members.

c. The department hires many of its undergraduates as lab and recitation T.A.s, research assistants, and computer assistants.

d. The students remarked that faculty members were willing to talk to them about anything, including questions dealing with materials in courses that the faculty member was not teaching.

Comments: A nurturing climate is the energy that makes the other elements of the framework successful. The success in implementing the framework is greater than just the sum of the elements. If the framework is to be successful, then there must be a genuine conviction and belief that the elements in the framework will make a difference in increasing the number of students who choose physics as a major. As one departmental faculty member stated, "warm fuzzes are not enough."

Element 14: Thriving physics departments have strong leadership.

Evidence: Although the word leadership may not have been used specifically within each department, department chairs and other administrators made strong inferences about leadership. Strong leadership was a key factor in assisting departments in changing the environment to attract a significant number of physics majors. In the majority of departments, most people were able to identify a specific person as a leader. This gave the impression of strong leadership through the change process.

C. Additional Concerns for Creating Thriving Physics Departments

After the review of the reports, it should be noted that there were several critical issues that were not addressed. Further, it seems appropriate for these issues to be of concern to NTFUP. Specifically, three (3) issues have been identified: (1) the preparation of high school physics teachers; (2) the educational preparation of women and minorities; and (3) the departmentalization of the elements for producing thriving physics departments.

First, there was little or no mention of interest in the preparation of high school physics teachers included in the reports. Of the 21 physics departments visited, 11 reported having some type of teacher education program; however, the comments about these programs made it clear that the preparation of high school physics teachers was not a focus for most physics departments. Two of the representative comments are as follows:

[The] physics major is a very rigorous degree program and it is very difficult for students to major in physics if they are interested in teaching at the secondary level (although an occasional physics major does go into teaching).

The department also has a Physics Teaching program that attracts one student per year. In addition, a few physics graduates join the Teach for America Program each year.

After reviewing the comments about the preparation of physics teachers, it seemed questionable about whether or not the maintenance and portrayal of the academic rigors of the major was at the expense of preparing people to use physics. It is also possible that the attitude about preparing high school physics teachers manifested the fear of lowering the standards of the major. If the number of physics majors is to increase at the university level, it seems unreasonable to expect that high school physics teachers can/will be able to prepare large numbers of potential majors if physicists are unwilling to teach the necessary content. Perhaps the physics community has recognized its negligence in preparing high school physics teachers because the NTFUP is piloting a nationwide PhysTEC Program, a very noteworthy effort for preparing high school physics teachers.

Although it might be issues of commitment, encouragement, and desire to do so, it should be pointed out that some departments already have the structure for producing high school physics teachers in place. The Masters of Arts in Physics Education (MAPE) that is offered by one of the departments already serves as a model program for assisting physics teachers who are already in the field.

The MAPE degree is designed to provide middle school physical science and high school physics teachers with a strong background in physics. The degree is designed for teachers who do not have an undergraduate degree in physics.

The physics education research and teacher preparation efforts of two other departments are also notable. But overall, only a few departments are showing concern about the preparation of physics teachers.

A second issue is that there was only an occasional mention of minorities and women in the departments, obviously not a strong point of emphasis. Again there are a few schools making a special effort to address this issue. Often when the issue of minorities is mentioned, it is assumed that the main concern is an issue of civil rights. To the contrary, this concern is for the future of physics and the acceptance that the perspectives of any given academic profession cannot rest solely with only one subgroup of the population—white males.

A related concern is the consideration of the white male subgroup of our society that the majority of physics majors come from, and whether this subgroup is increasing or decreasing in respect to its interest in physics. Due to the capabilities of the students in the present targeted pool, it is probable that they are excellent students not only in physics, but in other areas as well. Consequently, they will be attracted to other areas of study as well as recruited into other fields. If in fact potential physics majors primarily come from the subgroup of white males, then it is probable and conservatively estimated that 65% of the American population is not considered to be in the pool of potential physics majors. Further, if the U.S. demographic data continue as predicted, then the relative size of the pool of majors will continue to decline.

In one sense, finding a solution for the concern of increasing the presence of minorities and women in physics is related to the issue of preparing more and better high school physics teachers. More quality physics teachers in high schools will lead more potential physics majors to universities. However, finding a workable solution to the issue will be decidedly different from the issue of just preparing teachers. There are many sociocultural factors and barriers that must be overcome. The same elements of the framework will work for women and minorities, but there must be an understanding of cultural differences. More importantly, however, genuine attitudes of acceptance will remove barriers to learning. Credit must be given to NTFUP for its quick response in arranging a meeting scheduled for the fall of 2002 to begin addressing issues of diversity.

The third concern is a lack of departmentalization of the framework elements by many physics departments. The framework must become embedded within departmental policies, and the responsibilities for implementation must be shared by all faculty members. Although the term departmentalization was not specifically used in the reports, there was concern for the sustainability of practices once key people retired. This supports the concern that the framework needs to be departmentalized.

Summary

Concerned about the decline in the number of physics majors, members of the National Task Force on Undergraduate Physics (NTFUP) conducted a study to determine the causes of the decline and to determine strategies for reversing this phenomenon. Since not all physics departments have a declining enrollment in physics majors, NTFUP identified 21 thriving departments for study. To conduct the study, teams consisting of three physicists each visited the 21 physics departments for about one and a half days of observation. Each team summarized its observations and drafted a report. All reports were then examined to ascertain if there were any common themes that could possibly serve as a framework of elements for other departments to emulate. A framework consisting of fourteen (14) elements was identified as common to all thriving departments. Based on these elements, it appears as though there are practices that other physics departments can follow to increase the enrollment of physics majors.

There were also three concerns that might influence potential physics majors which were not mentioned in the reports but should be of concern to NTFUP. These issues are the lack of interest and focus in preparing high school physics teachers from many of the departments; the lack of minority and women representation in the departments; and the lack of departmentalization of the elements for producing thriving physics departments.

NTFUP is to be commended for its ability to take action by supporting PhysTEC, a nationally piloted program to increase the number of physics teachers and improve the preparation of high school physics teachers. Plans have already been made to begin addressing issues of diversity with a meeting for the fall of 2002.

Appendix VII.
Survey Form

The survey was administered as a web-based form. The following is a text version of the form.

National Task Force on Undergraduate Physics

Project SPIN-UP: Strategic Programs for Innovations in Undergraduate Physics Sponsored by the ExxonMobil Foundation

SURVEY OF UNDERGRADUATE PHYSICS PROGRAMS

1. Your name: _____

   Your title: _____

   Your institution: _____

   Email: _____

Physics courses and curricula

2. How many <u>total</u> credits (not only physics) are required to earn a bachelor's degree at your institution? __ __ __ credits

Is your academic calendar divided into (check one)

___semesters?   ___quarters?   ___other?

What does "one credit" typically represent at your institution?

   ___ One hour per week in class

   ___ One course for an entire academic term

   ___ One course for an entire academic year

   ___ Other (please describe)_____

3. Please respond below with information about your "most rigorous" physics program. This is usually the undergraduate curriculum that requires the largest number of physics credits and is often designed for students preparing for graduate study in physics. For this survey, we will refer to this program as your "standard" physics curriculum.

Degree title (check only one that gives the closest match with your "standard" program):

   ___ Bachelor of Science in physics

   ___ Bachelor of Arts in physics

   ___ Bachelor's in engineering physics

   ___ Bachelor's in applied physics

   ___ Other_____(please describe)

3a. How many credits are required in the following areas for this "standard" physics degree?:

   _____ Physics credits

_____ Mathematics credits

_____ Chemistry credits

3b. Does this degree track **require** (check as many as apply):

___a research experience?   ___a thesis?

3c. Total number of graduates (in this specific "standard" degree program) in past 3 years: _____ (graduation years 1999, 2000, 2001)

4. Indicate below **the number of credits** in each area required for your "standard" degree program described in Question 3. The total should equal the number of physics credits you entered as a value in Question 3a.

_____ Introductory classical physics (including lab, if appropriate)

_____ Introductory modern physics

_____ Intermediate classical mechanics

_____ Intermediate electromagnetism

_____ Mathematical physics

_____ Optics

_____ Thermal and/or statistical physics

_____ Quantum mechanics

_____ Advanced laboratory courses (including electronics)

_____ Other physics courses

5. **In addition to** the "standard" degree program described in Questions 3 and 4, which of the following alternative degree tracks do you offer? Check the tracks at left, and also indicate the required number of physics credits and the total number of graduates in that track in the past 3 years (graduation years 1999, 2000, and 2001).

Below: First column: Number of physics credits required

Second column: total number of graduates in the past 3 years

___ Bachelor of arts
    (only if "standard" degree is *not* B.A.)   _____   _____

___ Engineering physics   _____   _____

___ Applied physics   _____   _____

___ Physics degree for teachers   _____   _____

___ Specialized degree or concentration   _____   _____
    (e.g., geophysics, biophysics, computational physics etc.)

___ Combined degree (e.g., physics + math,   _____   _____
    physics + business, etc.)

___ Astronomy degree (if offered through a   _____   _____

    separate department check here___)

___ Other_____   _____   _____

_____   _____   _____

___ None of the above

5a. Are you planning to add any alternative degree tracks in the near future? Please describe._____

_____

_____

6. Does your department or program offer a minor?

    __ No   __ Yes, in physics   __ Yes, in astronomy

6a. How many **physics** credits are required for the **physics** minor? ___ credits

6b. How many students minored in physics over the past 3 years? ___

6c. How many students minored in astronomy over the past 3 years? ___

7. Averaged over the last 3 years, approximately **how many** of your graduating seniors participated in the following activities each year:

    ___ Undergraduate research on campus

    ___ Undergraduate research off campus (such as REU or industrial internship)

    ___ Presented research results at local, regional, or national meeting

Recruiting physics majors

8. Which, if any, of the recruiting activities below does your department pursue? Please check all that apply:

Recruiting of high school students:

    ___ Hold annual (or more often) departmental open house for students & parents

    ___ Host individual prospective students and their families in the department

    ___ Hold summer workshops for high school students

    ___ Faculty or students regularly visit local schools

    ___ Target recruitment of students likely to major in physics

    ___ Target recruitment of students who are underrepresented minorities

Recruiting of enrolled college students:

    ___ Identify and recruit talented students in service courses

    ___ Offer "introduction to the profession" courses for first-year students

___ Group potential physics majors in special section of the introductory course

___ Actively recruit transfer students from two-year colleges

___ Other (please describe) _____

8b. Please describe the recruitment activity or activities that you consider most successful in attracting majors to your department:_____

_____

_____

Interactions between physics faculty and physics majors

9. Who has **primary** responsibility for advising physics majors?

___ Several or all physics faculty members

___ One physics faculty member (other than the department chair)

___ The department chair

___ A non-faculty physics department staff member

___ University advisors outside the physics department

___ Other (please describe) _____

10. On average, how often do most physics majors interact with their advisor(s)?

___ Less than once per year

___ Once per year

___ Once per semester or quarter

___ Several times per semester or quarter

11. Which of the following does your department do for students? (*Check all that apply*)

___ Assign a faculty mentor to each student

___ Assign a peer mentor to each student

___ Provide dedicated undergraduate study room or lounge

___ Have an active physics club or SPS chapter

___ Provide building keys to undergraduate physics majors

___ Conduct exit interviews with graduating seniors

___ Other activities that enhance undergraduate program (*please describe*)

_____

11b. Which (if any) of these activities does your department consider most effective in shaping student attitudes regarding your department? Please explain.

_____

_____

12. Which of the following have you used in the past year to provide career information to your undergraduates? (*check all that apply*)

___ Alumni visits to the department

___ Field trips to local industries

___ The university career services office

___ Departmental colloquia by physicists in industry

___ Career materials from the professional societies

___ Other (please explain) _____

___ None of the above

12b. Which of these activities (if any) does your department find to be most useful in guiding your students' career choices?_____

_____

_____

Alumni of the Department

13. What percentage of your alumni from the past three years have gone into the following areas?

Graduate study in physics	___ %
Graduate study in other area	___ %
Employment in technical field	___%
Employment in nontechnical field	___%
High school teaching	___%
Continued in a 3/2 engineering program	___%
Other	___%
Don't know	___%

14. What type of information does your department currently maintain on its alumni? (*Check all that apply.*)

___ Updates from past students by email or phone

___ Mailing or email addresses for students at the time they graduate

___ Information on employment or graduate school plans at time of graduation

___ Mailing list for departmental newsletter

___ Surveys of alumni

___ Other (please describe) _____

___ None of the above

Curricular reform

15. Have you made significant changes in your curriculum over the last several years?

    ___ Yes (*if yes, please continue to question 16*)

    ___ No (*if no changes were made, please skip to question 19*)

16. For each area in which changes were made, please specify whether the changes were made in content or in the way in which the courses are taught (pedagogy).

	Content	Pedagogy	Both	N/A
General education courses:	___	___	___	___
Algebra-based introductory course:	___	___	___	___
Calculus-based introductory course:	___	___	___	___
Introductory course for majors:	___	___	___	___
Upper-division courses:	___	___	___	___

17a. What motivated or prompted the curriculum changes? (e.g., threats from dean, energetic individual faculty member, external review committee, complaints from students, etc.) Please explain.

_____

_____

_____

_____

17b. What were these changes intended to accomplish? (e.g., increase introductory enrollments, increase number of majors, improve preparation of students for graduate school or careers, etc.) Please explain.

_____

_____

_____

_____

17c. What measures or indicators of success do you have for these changes? (If none, please state "none.")

_____

_____

_____

_____

18. How were the costs of these changes financed? (*Check all that apply.*)

___ Internal reallocation of resources within the department

___ University or endowment funds from outside the department

___ Grant(s) from private foundation(s)

___ Grant(s) from NSF or other federal agency

___ Funds or equipment from business or industry

___ Other (describe)_____

19. What are your undergraduate program's greatest strengths?

_____
_____
_____

20. What are your undergraduate program's greatest needs or challenges?

_____
_____
_____
_____

Appendix VIII.
Case Study Documents

The following pages contain the case study documents summarizing the findings of each of the 21 site visits. Each document is based on the report written by the site visit team. The chairs of the site visit departments have checked the documents for factual accuracy. The opinions expressed in the documents are those of the site visit teams.

The pdf version of this report, available through the American Association of Physics Teachers website (at http://www.aapt.org/Projects/ntfup/casestudies.cfm) has the case study documents with color pictures accompanying each document.

A. CASE STUDY: Angelo State University

The Setting:

Regional comprehensive university of 6,300 students. A majority of students come from the agricultural area around San Angelo, TX, a city of 100,000 where the university is located.

Enrollment is 25% minority (19% Hispanic, 5% Black, 1% Asian, and 0.3% Native American).

Department of Physics has eight full-time faculty (including a planetarium director and a geologist), two part-time faculty, and a departmental secretary. Student help is limited to a student tutor, a work-study student to help the secretary, and four students per semester to help with grading in large sections.

What Has Been Done:

1. The department offers three tracks to a bachelor's of science degree in physics as well as pre-engineering tracks in conjunction with other Texas universities.

    – The three major programs are physics, applied physics and physics with secondary certification. The department offers 3/2 Physics Engineering programs through articulation agreements with engineering departments at Texas A&M University, the University of Texas at El Paso and Lamar University. In addition, they have a 4/2 program leading to a masters degree in Semiconductor Physics or Semiconductor Engineering in conjunction with Texas Tech University.

2. The department has made a conscious decision not to pursue a graduate program in physics but to concentrate its efforts on undergraduate education.

    – The department has a unified sense of mission and focuses on high-quality education in small classes, and recruiting talented students. A particular focus is the preparation of students for careers in the health sciences.

3. The department cultivates a strong sense of community between students and faculty as well as within the student and faculty communities.

    – The SPS chapter is active and well-mentored. Entering students are encouraged to seek out upper classmen for mentoring and to join SPS. Students find the faculty accessible and perceive small classes as a great benefit of being at Angelo State. Students are advised on career goals and academic programs during pre-enrollment visits to campus, freshman orientation and as a part of regular advising periods. Each student is assigned a faculty advisor when he or she enters the university, and students cannot enroll for courses without permission from their advisors.

4. Student research is an important component of the ASU program and is characterized by the department as "the current capstone experience for majors."

    – Every physics major is required to complete a three-hour research course prior to the fall semester of their senior year, and the department typically has three to five students registered for undergraduate research per semester. More than half the faculty are engaged in research and the department has received more than $100,000 in funding to support undergraduate research. The Carr Research Fellowship, an internal funding source, also supports one to three students each year, and other students complete the research requirement through summer REU experiences.

5. The department has increasingly worked to recruit physics majors both in local high schools and in service courses.

   – The department makes annual recruiting trips to regional secondary schools, organizes on-campus programs for secondary students, and participates in college recruiting efforts. The department has several scholarships that are available to attract talented students.

Indicators of Success:

1. During the past 10 years, the department has graduated an average of eight students annually.

2. Approximately 1,400 students per semester enroll in physics, physical science, astronomy or geology giving the department the highest class size average (44 students) in the College of Science.

3. Several physics majors have recently completed medical school at prestigious institutions.

4. There appears to be a strong high school physics teacher bond and loyalty to Angelo State in recommending it to students.

Keys to Making the Changes:

1 The department benefited from long-term stability of faculty from 1960–1990 including both the Vice President for Academic Affairs and the President of the university as part-time faculty. Ten years ago, these faculty set about slowly and steadily hiring replacements who would be sensitive to the needs of Angelo State's students.

2. Department chairs have provided strong leadership in recruiting new faculty interested in teaching and research without fragmenting the strong student faculty cohesion that is a strength of the department.

3. The department works collegially with an unusually strong sense of shared mission. This shared mission is helped by the relative isolation of the university.

For More Information Contact:

Andrew Wallace, Head, Department of Physics, Angelo State University, PO Box 10904, 2601 West Ave. N, San Angelo, TX 76909; Phone: 915-942-2242; Email: awallace@angelo.edu.

B. CASE STUDY: University of Arizona

The Setting:

A public, land grant, Research 1 institution with about 25,000 undergraduate students. The Department of Physics is administratively within the College of Science, which enrolls about 3,000 majors. Other physics-related programs in the College of Science include Astronomy, Atmospheric Sciences, and Planetary Sciences. The Optical Sciences Center is a world-renowned facility that independently offers undergraduate and graduate degrees in optics.

The Department of Physics includes 28 professorial faculty (seven assistant professors, one associate professor, 20 full professors) and two lecturers. The unusual distribution between assistant and full professors is a result of a number of retirements and an aggressive recruitment program to bring in new faculty. Major research programs in physics include AMO physics, astrophysics, biophysics, condensed matter physics, nuclear physics, elementary particle physics, and physics education. Two staff members have responsibility for the lecture demonstrations and the instructional laboratories.

Beginning in the 1960s and lasting through the 1980s, the university concentrated on building its reputation as a Research 1 institution and so focused its efforts on developing its research and graduate programs. In the 1990s, the Department of Physics, with strong support from the college and university administrations, undertook an ambitious program to strengthen its undergraduate program.

What Has Been Done:

1. In addition to its two traditional undergraduate degree programs (the standard physics B.S. degree and an Engineering Physics degree offered through the College of Engineering and Mines), the department has developed a B.A. degree and a B.S. in Science Education. The B.A. is intended for students who are interested in physics but not for professional careers (pre-law students, for example).

2. A separate option within the B.S. degree allows students to substitute courses in atmospheric science for certain physics courses and thereby achieve an interdisciplinary degree. Other such options are under consideration.

3. A diverse array of introductory courses is available, including those that serve biological science majors (both algebra-trig and calculus-based courses are available), engineering, and science majors. A separate four-semester course is offered for physics and astronomy majors.

4. A joint physics degree program for astronomy majors has been developed. This has increased the total number of physics graduates and has been especially effective in increasing the number of women physics graduates (because women generally constitute a higher fraction of astronomy majors than physics majors).

5. The department and the college have embarked on an ambitious program to train science teachers within the College of Science. Four new faculty positions were created (one in physics). A common and jointly taught set of teacher education courses has been developed.

6. There is an active program of reform in the introductory courses. Such techniques as peer instruction, interactive lecture demonstrations, and inquiry-based laboratories are being used.

7. The department maintains an exemplary outreach program, which includes enrichment activities for pre-college students and contacts with alumni such as a professional-

looking departmental magazine and a program that brings alumni back to the department for talks.

Indicators of Success:

1. The number of physics degrees awarded has remained fairly steady at about 22 per year despite the national decline. In addition an average of about six Engineering Physics degrees are awarded each year. About 25% of the degrees are awarded to women.

2. Even though many first-year students switch to other majors, the decrease is mitigated by a large number of students who transfer into physics or who enroll as double majors in the junior or senior year.

3. The department has been selected as one of the six initial sites to develop the PhysTEC program, which has been funded by NSF and the U.S. Department of Education.

4. Students appear to be very satisfied with the department and its programs and faculty.

Keys to Making the Changes:

1. The dean of the College of Science is very supportive of the department's efforts in undergraduate education and intends to support hiring to bring the department back to about 40 faculty.

2. The department head has made a significant commitment to improving the undergraduate program and has provided strong administrative support for faculty who are involved with the course improvements.

3. An active SPS chapter provides a unifying element for students. The department has provided the SPS with a laboratory room where students can work on their own to gain experimental skills and learn about the operation of laboratory equipment.

4. The department makes an effort to be open and accessible to students. The department head is accessible to students in his office and holds periodic "town hall" meetings with undergraduates. Students are asked to serve on departmental committees. The department has instituted a greater level of recognition for students who excel in their academic work.

5. A Science Education program has been developed within the College of Science to prepare K–12 science teachers; this program has stimulated reforms in the introductory courses.

6. The department involves many of its students in the research program and as undergraduate T.A.s.

7. An Academic Support Office in the Department of Physics helps students with schedule changes, arranges advising and tutoring, and provides information on scholarships and summer research opportunities. An employment database provides information on internships as well as on permanent positions.

8. Newly hired faculty are enthusiastic about the curriculum reform efforts. There is an active mentoring program for new faculty. A departmental teaching award recognizes outstanding accomplishments in teaching.

9. The department has funded a staff position as Communications Director to support the outreach programs.

For More Information Contact:
Daniel L. Stein, Head, Physics Department, University of Arizona, Tucson, AZ 85721; Phone: 520-621-4190; Email: dls@physics.arizona.edu.

C. CASE STUDY: Bethel College

The Setting:

Liberal arts college enrolling about 2,500 students founded by Swedish Baptists and retaining a strong Christian tradition. Enrollments are increasing and admission standards are gradually rising. The college accepts federal funding for faculty research but does not accept federal funding for activities solely directed towards instruction.

The department has recently expanded to four tenure track faculty members, has one adjunct faculty member and no staff although they use undergraduates as T.A.s for tutoring and lab assistance. Sixty majors enrolled in four physics tracks and a 3/2 program with an average of 12 graduates each year.

About a third of the graduates go to graduate school. Another third participate in the 3/2 engineering program; 8% become high school teachers, and the remaining quarter go directly into the workplace. Over the last decade, Bethel physics graduates have earned 16 Ph.D.s and another eight are currently Ph.D. candidates.

What Has Been Done:

1. The department is student-oriented, and faculty work to ensure that students are members of the physics family and know they are.

   – The introductory course is carefully taught to introduce students to the department. Majors from that class are recruited as T.A.s in their sophomore year so they know the next year's freshmen. Some upper-division courses are taught every other year so juniors and seniors work together.

   – The department provides keys to the labs to all majors, and any student who wants one has a desk in the labs or stockroom. Faculty are often in their offices nearby, even at night.

   – Each declared major is assigned a faculty advisor with whom they must interact at least twice a year. Any of the faculty can act as an advisor, and all work to keep students informed of opportunities like REU programs and internships.

2. The department encourages students' involvement in research. A research experience is required for the B.S. degree in physics.

   – Three weeks of the laboratory in each semester of the introductory course are devoted to student projects. With faculty guidance, student groups of three select projects, design and conduct the experiments, and prepare oral and written reports on them.

   – The optics course serves as an introduction to research in which students complete projects that are sophisticated enough for presentation in regional meetings and frequently serve as the basis for senior projects. Throughout the course, students are required to observe the forms of physics publications including preparing papers in laTEX with abstracts, data tables, graphs and photographs.

   – Faculty hire students in the summer as research assistants and encourage them to participate in REU experiences and internships in local industries.

3. The department accommodates the needs of its students by offering a B.S. in physics, a more flexible B.A. in physics, a major in Physics Education for preservice high school teachers, and a new B.S. in Applied Physics for students who plan to seek jobs in

industry or pursue graduate degrees in engineering or other interdisciplinary areas. They also have a 3/2 program in engineering in which about a third of their graduates are enrolled.

4. The faculty of the physics department have and work to maintain very close ties with local industries. Local industries employ students as interns and frequently hire graduates. Faculty work as consultants and have received support for research projects conducted at Bethel with student collaborators. Industries have donated state-of-the-art equipment to the department.

5. The department works to recruit talented high school students to Bethel through such means as hand-written letters and phone calls from the chair, presentations at local high schools, participation in summer programs for gifted students, and awarding of departmental scholarships.

6. The department has recruited talented retired high school teachers to design and teach a course for preservice elementary teachers. The course actively involves the teachers in doing physics and has been very popular and successful.

Indicators of Success:

1. The department graduates an average of 12 majors per year.

2. The department added a new tenure-line faculty position this year.

3. Since 1998, the faculty have published or presented 71 papers with 22 student coauthors.

4. About 10 students each year are employed as interns by local industries.

5. The college administrators consider the Department of Physics as a showcase department.

6. The students and faculty clearly work as colleagues, and alumni appear happy with the education they received at Bethel.

Keys to Making the Changes:

1. The faculty work extremely hard and are dedicated to their students and the college. The students reflect the faculty's work ethic.

2. The college administrators strongly support the work on the Department of Physics with both resources and faculty positions.

3. The Department of Physics has adequate funding for its programs from annual supply and equipment budgets, a $55 lab fee charged by the college, an endowment from a former professor, and outside research funding.

4. The department has worked to establish connections to the national and regional physics community both to improve the department's professional reputation and to provide their students with opportunities to participate in physics meetings.

5. Bethel College works to promote the growth of faculty members as scholars and teachers by careful mentoring and a five-year tenure review. This results in a very low faculty turnover, and the Department of Physics has benefited from this stability.

For More Information Contact:

Brian Beecken, Chairman, Department of Physics, Bethel College,
3900 Bethel Drive, St. Paul, MN 55112; Phone: 651-638-6334;
Email: b-beecken@bethel.edu.

D. CASE STUDY: Brigham Young University

The Setting:

A private, church-sponsored university with 32,000 students, 30,000 undergraduates. Ninety-eight percent of students are members of The Church of Jesus Christ of Latter-Day Saints; 50% are women and 10% are under-represented minorities. Most male students go on a two-year church mission after their freshman or sophomore years and are more mature than typical college juniors.

The Department of Physics and Astronomy has 28 full-time faculty [22 on Continuing Faculty Status (tenured)], two research associate professors, 10 full-time staff, and two staff who support commitments to the *X-Ray* journal and the Astronomical Society of the Pacific.

Teaching loads are 15 credit-hours per year and the department currently has about 25 graduate students. The department currently has almost 300 undergraduate physics and astronomy majors. Of current undergraduate majors, 46% are majoring in physics; 21% in astronomy; 14% in applied physics–selected options; 12% in applied physics–computer science; and 8% in physics teaching.

What Has Been Done:

1. For the past five years, all physics majors have been required to complete a senior thesis, honors thesis, capstone project or teaching experience before graduation.

    – Faculty mentor every student in one of these experiences. Two-thirds of the faculty advise undergraduate research projects which involve well over 100 student projects each semester. The university provides $20,000 per semester to support 20–25 undergraduate research students and the department and external grant funds support another 10 or so. REU funds support 10 undergraduates each summer although not all of them come from BYU.

2. The department has broadened and provided more flexibility in its degree requirements.

    – The department offers a standard physics major, a bachelors of arts for preservice teachers and an applied physics option. The applied physics option allows students to select courses from outside the department in areas such as other sciences, engineering, law, business or computer science and have them approved by the chair to complete their degrees.

    – The department as a whole has not adopted a single pedagogical style. However, faculty are aware of the results of Physics Education Research and have changed the way they teach in such ways as introducing more conceptual questions or using group work.

3. The department has expanded its advisement program.

    – The department has made advisors more available, hired student advisors and put them in highly visible locations, instituted an introductory seminar course, and improved the student handbook so that its style is welcoming to students and its content valuable and useful. Majors are encouraged to meet frequently with their advisors, and there is an on-call faculty advisor. The associate chair meets with students having difficulty to work out a plan to improve their performance. There are an undergraduate coordinator, an undergraduate research coordinator, a capstone project coordinator and a physics teaching coordinator who advise majors. Finally, research students are "adopted" into the group and get advice from the rest of the group. The system is well-publicized.

4. The department provides a nurturing environment for majors.

    – Faculty members are readily available for student questions during their time on campus. They treat all students with respect and dignity. The SPS chapter has become much more active drawing 40–100 students at activities, and there is an active astronomy club. All newly declared physics majors take a 1/2-credit hour course on "Introduction to Physics Careers and Research," and students returning to school after their mission or another break are offered a one-credit hour course reviewing physics and math. The SPS space is integrated with the tutorial area which also houses the peer advisors. The department provides many non-research employment opportunities for students. The department maintains a "family" of alumni and has close ties with local schools.

5. The department views preservice and in-service education for K–12 teachers as an important part of its mission.

    – Faculty are working to involve students enrolled in the B.A. Physics Education option in more day-to-day activities in the department. They have a RET grant that supports four teachers each summer and are seeking funding for further activities.

Indicators of Success:

1. The department graduates 45–49 physics majors each year for the last three years, up from about 36 in the recent past. There is a high retention rate among majors, and students switch into physics from related disciplines such as math and engineering.

2. The number of freshmen physics majors has nearly doubled over the last 10 years; entering majors are increasingly better students and better prepared.

3. The percentage of female physics majors has increased from 19% between 1991–1996 to 26% over the last five years.

4. Ten percent of graduates enter teaching.

5. More than half of all graduates give papers on their research at regional or national meetings or conferences.

6. Graduates are accepted into quality graduate programs and recruited for jobs. Surveys of majors, alumni and graduating seniors reflect positively on the physics major program.

Keys to Making the Changes:

1. All faculty members are strongly committed to teaching and the excellence of the undergraduate program. The department and the university support this commitment with faculty time and dollars.

2. Changes in the undergraduate program come from suggestions made to or by the Department Undergraduate Committee and are presented to the faculty in writing and discussed and voted on by the entire faculty. The decision is usually by consensus as the department operated in a collegial manner. Faculty who suggested change usually volunteer to implement it.

3. Faculty have a high degree of collegiality and willingly collaborate in areas such as teaching, sharing equipment and sharing expertise. There is a sense of shared mission in the department.

4. The university administration tries to provide "top down support for bottom up ideas" and has been very supportive of the physics department's ideas for improving its undergraduate program in terms of resources as well as of moral support. Undergraduate research is one of the highest priorities of university fundraising.

5. An enrollment cap imposed by the Board of Trustees has increased the academic qualifications of entering students which has facilitated reform of the undergraduate physics program.

6. The sponsoring church contributes a unique environment to BYU including financial stability, support for scholarship by students and faculty, an unusual student body with common values and a strictly enforced honor code, as well as an edge in recruiting students who are members of the Church of Jesus Christ of Latter-Day Saints.

For More Information Contact:

Steven Turley, Chair, Department of Physics and Astronomy, Brigham Young University, NA281A ESC, Provo, UT 84602-4679; Phone: 801-378-3095; Email: turley@byu.edu.

E. CASE STUDY: Bryn Mawr College

The Setting:

Highly selective, liberal arts women's college with 1,200 undergraduates and 400 graduate students. It attracts very talented students who are culturally and geographically diverse and offers financial aid to more than half. 20% of students are of Asian heritage with a smattering of other minorities and foreign students.

The Physics Department has four tenure-line faculty members, a full-time laboratory coordinator who holds a Ph.D. in physics, two lecturers, one of whom has been at the college for four years and the second of whom is a sabbatical replacement, and a full-time departmental secretary. The department has access to a fully equipped machine shop, a library and state-of-the-art computer facilities. Teaching loads are five courses per year, and all tenure-line faculty are active in research.

The Physics Department at Bryn Mawr has a reciprocity agreement with nearby Haverford College. The two departments take turns offering upper-division physics courses which are open to students at both institutions. The department also has a small Ph.D. program that currently has four students enrolled. These students fulfill course requirements either as reading courses with Bryn Mawr faculty or by agreement with the University of Pennsylvania.

What Has Been Done:

1. The department works to create and maintain a family-like culture within the department. Their students are women, and the faculty work hard to build students' confidence in their ability to succeed in physics.

    – Students and faculty members are on a first-name basis. Student representatives attend faculty meetings and have a voice in departmental governance. The faculty work hard for the success of their students whether or not these students plan to attend graduate school in physics.

    – Physics majors are given building keys and keys to a "majors common room" and a computer lab. There is a kitchenette in the department, and students use numerous other rooms as study spaces. Students are provided with a handbook that is updated every year. All upper division majors have mail boxes in the department.

2. The major is designed to be flexible enough to accommodate the career goals of a variety of students. In addition, students who start physics in the fall of their sophomore year can complete a physics major in four years.

    – Two years of the major are completed at Bryn Mawr before students start upper division work, some of which is done at Haverford where most physics majors are male. This is a conscious decision to allow women majors time to gain confidence in their choice of profession.

    – Bryn Mawr physics courses are lab-intensive, and students are encouraged to work cooperatively as well as alone in order to maximize their confidence in their ability to manipulate equipment.

3. The department actively recruits as physics majors students enrolled in the introductory course. Faculty take considerable pains to encourage all students in the course and to point out to talented students that they could succeed as physics majors and the advantages that such a major offers.

4. The department stresses careful advising that works with students on all aspects of their

lives. One faculty member is assigned as chief advisor to students in one class and gets to know each student well. Faculty advisors actively seek students to be sure that they are encouraged to move forward in their chosen careers.

   – The department is aware that only 40% of their majors will pursue a Ph.D. in physics and carefully advise the 29% who go directly into the workplace, the 9% who enter teaching, the 16% who plan to enter professional school, and the 7% who pursue masters degrees. The faculty track their alumni and are proud of their diverse careers.

5. Students are encouraged to participate in research. They are given tours of faculty labs in introductory courses and majors are invited to work with faculty in their labs. Because the faculty is small, students are also encouraged to participate in REU experiences at other universities, or take internships in industry or at national labs. A senior thesis is not a requirement for a degree, but students, particularly those planning to attend graduate school, are encouraged to complete one.

Indicators of Success:

1. The department graduates an average of 10 physics majors per year.
2. Morale among students is very high. They are proud of succeeding in a demanding curriculum.
3. Bryn Mawr is near the top of any list of colleges and universities in terms of production of numbers of women physics majors.
4. In recent years, 1–2% of the 150 women earning Ph.D.s in physics each year received their bachelor's degree at Bryn Mawr.
5. Nonmajors enrolled in service courses appreciate the supportive nature of the department, although they are not as enthusiastic as the majors about the laboratory experiences.

Keys to Making the Changes:

1. The faculty care deeply about their students and take their teaching role very seriously.
2. New faculty members immediately become members of the departmental team and are nurtured by senior faculty members.
3. The department works by consensus decisions of all faculty members reached through discussion at weekly faculty meetings which may include student representatives.
4. Faculty rotate administrative duties so government by consensus is very important.
5. The college administration has encouraged increasing the numbers of majors in disciplines where women are under-represented. The Admissions Office has actively recruited students who are interested in and capable of succeeding in science and mathematics, and enrollments in the sciences have increased across the board by 50% (except in chemistry).
6. Entering students generally have higher math SAT scores and lower verbal scores than students of 20 years ago, and there are currently a significant number of students of Asian background.

For More Information Contact:

Elizabeth McCormack, Chair, Physics Department, Park Science Center, Bryn Mawr College, 101 North Merion Ave., Bryn Mawr, PA 19010; Phone: 610-526-5356; Email: emccorm@brynmawr.edu.

F. CASE STUDY: California Polytechnic State University at San Luis Obispo

The Setting:

A comprehensive, four-year public university with 18,000 students whose philosophy is "learn by doing." Rated as the best public, largely undergraduate university in the West for the last nine years. One of three founding programs, the engineering program, has more than 4,500 majors.

The Physics Department of the College of Science and Mathematics has 30 faculty members, three technical staff members and two administrative staff. The department offers courses in astronomy, geology, geophysics, and oceanography as well as physics and offers B.A. and B.S. degrees in physics and the B.S. in physical science. The department graduates 15 majors per year of whom 25–35% go on to graduate school.

What Has Been Done:

1. The department has a tradition of being friendly and open with a broad, hands-on and can-do approach to physics.

    – The physics program is designed to be flexible, preparing students for graduate study in physics or related fields, a teacher-credentialling program, or work in industry. Students have the opportunity to participate in a wide variety of research programs.

    – Once students identify themselves as physics majors, they receive phone calls from a faculty member welcoming them into the department community and offering information and advice. The office staff actively helps physics majors in matters of enrollment and other academic details. The chair and most of the faculty are aware of the student community and actively foster it.

2. The department requires completion of a senior research project for all degrees.

    – A faculty member is assigned to help students formulate a project in one of the areas where the department has strength. The projects must involve creative and original efforts by the student and include traditional research projects as well as other types of projects.

3. The department has invested in a studio-style teaching laboratory which is not yet used for the upper-level majors' courses but has had an impact on faculty and student T.A.s who are involved in the course and the pedagogy is "leaking" into other physics courses.

4. The department has created a student-operated and controlled lounge area known as the h-bar ($\hbar$) which they describe as the "beating heart" of the physics program.

    – The h-bar is centrally located in the department, near faculty offices and space used by seniors for their research projects and is large enough to accommodate a large group or several small ones. Seniors feel responsibility to help younger majors, and the majors develop an unusually high degree of cohesiveness and have formed a genuine learning community. The h-bar is a home away from home for students but remains strongly focused on the work of the discipline.

5. The department has fostered the active and direct involvement of the technical staff in the educational experience of the students.

    – Many students interact directly with the technical staff, particularly during their senior projects, and staff see helping students as a part of their jobs. Students are required to complete five quarters of lab experience which include projects that are student-initiated but completed under the guidance of an experienced technician. The friendly and respectful relationship between students and technical staff is a unique attribute of this program.

6. The requirements for retention, promotion and tenure are clearly and explicitly defined by the college in which the department resides. Younger faculty express appreciation of very clear guidelines, and this, in turn, provides an unusually stable environment for the department so that all faculty feel free to focus on the quality of instruction.

Indicators of Success:

1. The department graduates an average of 15 majors per year, a number that has recently been increasing.

2. Students perceive the faculty and staff of the department as friendlier and more open than other departments on campus.

3. There are a number of women active in the department and occupying leadership positions.

4. The selectivity of the department (and the university) and the quality of physics undergraduates has been increasing.

5. Graduates succeed in selective graduate schools and other careers.

Keys to Making the Changes:

1. Large enrollments in engineering offer a fertile recruiting ground for physics majors and the university has a tradition of strength in science and technology. The physics program has benefited as the university, helped by its high ratings in *U.S. News and World Report*, has attracted better prepared students.

2. There is an uncommon level of respect among all elements of the department: faculty, lecturers, administrative and technical staff and students.

3. Careful sequencing of classes fosters the development of student networking as does the provision of a student-controlled space in the department.

4. The pre-identification of majors encourages many physics majors to enter Cal Poly, and the fact that it is easy to identify majors in their first year allows mentoring of beginners by the department and fellow students.

5. The department and the administration encourage innovation in teaching from both the faculty and lecturers and the technical staff.

For More Information Contact:
Richard Saenz, Chair, Department of Physics, Cal Poly State
University, San Luis Obispo, CA 93407; Phone: 805-756-2447;
Email: rsaenz@calpoly.edu.

G. CASE STUDY: Carleton College

The Setting:

An academically competitive, nonsectarian, liberal arts college with 1,902 students approximately evenly divided between men and women. It has historic strength in the sciences and mathematics and leads the liberal arts colleges in physics and astronomy graduates who obtain Ph.D.s in these fields.

The Physics and Astronomy Department has seven tenure-line faculty members, an instrument maker, an administrative assistant, and an electronics and laboratory manager. The department also employs a fifth-year Educational Associate who helps with the observatory and other astronomy equipment. There are currently 44 junior and senior majors.

In 1995, the department noted a significant drop in the number of declared majors. From 1993 to 1996, the department averaged 19 graduates per year. From 1997 to 2000, the average was 10.5.

What Has Been Done:

1. Instructors recruit physics majors during the introductory courses which are always taught by experienced and student-oriented faculty members.

    – Students at Carleton declare majors in the spring of their sophomore years after taking a number of courses in physics as well as other disciplines that interest them. Thus the department relies on the college's very able admissions staff to recruit able students and provide potential physics majors with information about these programs. They focus recruiting on the introductory physics courses and use them to develop a personal connection between students and a faculty member.

    – Carleton students take three courses for each of three terms. Each course is worth six credits and is equivalent to a semester-long course. The introductory physics sequence has two entry points, a three-credit course on Newtonian mechanics or a three-credit course on applications of Newtonian mechanics in the planetary realm aimed at better prepared students. The second three-credit course focuses on special relativity and particle physics topics few students have seen and which they find fascinating. Students are able to "sample" physics without committing themselves to a full year course.

2. The curriculum for physics majors is flexible and intended to integrate with a liberal arts education and prepare students for a variety of future careers as well as graduate and professional study in physics and other fields.

    – The former six-credit Classical Mechanics course has been split into two 3-credit courses, Classical Mechanics and Computational Mechanics, in order increase students' training in computational physics.

    – A required course on Waves was replaced by an applied physics requirement in which students select from a list of courses. This adds flexibility to the curriculum and allows students to spend a term off campus without getting out of sequence with the major.

    – Students are required to complete a senior "integrative exercise" in which they work with a faculty advisor to prepare a major paper and make an extended oral presentation on a topic chosen to integrate the physics they have studied in various courses.

3. The department works hard to build a sense of community among students and faculty in the department.

    – Students have keys to the physics building and a lounge as well as access to lab space and computers after hours. Pictures of all majors are posted in the lounge. Students and faculty have one or more meals together each week and may include visitors or speakers in them. There are organized social events including a spring and fall departmental picnic and a senior canoe trip.

    – After 7:00 p.m., the physics building belongs to students and typically eight to 10 of them can be found. They regard their fellow physics majors as an important element in their education.

    – The departmental curriculum committee consists of six students and two faculty members. When the numbers of majors dropped, the faculty listened carefully to student opinions.

4. The department encourages students to pursue a variety of careers through a course on "What Physicists Do" which brings five speakers to campus in a term to discuss their use of physics in diverse careers.

5. Students are encouraged to become involved in research with faculty some of whom have outside funding and all of whom remain active in physics. The college has funding to support some students during summer research projects, and all faculty work on research during the summer.

Indicators of Success:

1. The department graduated 16 majors in 2001 and currently has 26 seniors and 18 juniors who have declared physics majors.

2. Faculty and students both express enthusiasm for the community they find in the physics department.

3. Student morale is very high. Students participate in departmental decisions and feel ownership of their own education.

4. The administration respects the work of the department and is well-informed about its activities.

Keys to Making Change:

1. All faculty consider teaching their first priority, and promotion and tenure are based primarily on work in the classroom and interactions with students.

2. There is a strong personal interaction among faculty and students that centers on individual learning, growth and development.

3. The department has a shared understanding of its mission and an intense sense of purpose in fulfilling it. Decisions to make changes are made after open discussion by consensus of the faculty.

4. Students participate actively in determining curriculum and changes that should be made in it.

    For More Information Contact:
    Richard Noer, Chair, Department of Physics, Carleton College, One North College St., Northfield, MN 55057; Phone: 507-646-4387; Email: rnoer@carleton.edu.

H. CASE STUDY: Colorado School of Mines

The Setting:

A public research university with 3,000 students of whom 2,500 are undergraduates. Eleven undergraduate degree programs have expanded from traditional emphasis on mining to include energy, mineral, materials science, engineering, and associated engineering and science fields. Twenty-five percent of students are women and 15% are underrepresented minorities.

The Department of Physics has 14 tenure-line faculty, four adjunct professors, one research professor and two faculty members who are on phased retirement. One faculty member serves as laboratory instructor and lecture demonstration coordinator. Another serves as a full-time lecturer with no research responsibilities. The department also has a computer system administrator, an administrative assistant, a machinist and an electronics technician. The department awards three M.S. degrees and three Ph.D.s per year.

The undergraduate physics program, engineering physics, is one of 14 ABET accredited physics programs in the country. In 1996, the department established a five-year program leading to a B.S. in engineering physics and a masters degree in mechanical, electrical or electronic materials engineering. 20% of the department's majors are involved in the five-year program. All undergraduates take calculus-based physics as part of the core curriculum. One-third of graduates attend graduate school, roughly half in physics and half in engineering fields; a third take jobs in industry; and the rest follow a variety of paths.

What Has Been Done:

1. In 1984, the department decided to focus its program on engineering physics.

    – Requirements in the upper-level curriculum were reduced to allow students to take more applied courses although traditional upper division courses are available to students who need them. The department focused its research efforts on applied areas in nuclear physics, condensed matter physics and optical physics.

2. The department actively recruits students by emphasizing the flexibility of its programs and the wide range of careers to which they lead.

    – They provide a list of recent graduates and their current employment. The department participates in university recruitment days for high school students and their parents. During the required freshman seminar, physics faculty and students man posters at Options Days. The five-year program is advertised during the second semester of the required physics course. The department head assigns its best teachers to the large introductory sections.

3. The department teaches all students in the university introductory physics in sections of 90–100 students. Although students view the second semester of this course as one of the most difficult in the university, the faculty use this opportunity to recruit talented students as engineering physics majors.

4. The department works to build a community among faculty and undergraduate students.

    – Undergraduate majors have building keys and access to the library, a computer area, and an informal "penthouse" with a microwave etc. Faculty maintain an open-door policy, and their labs and offices are close to classrooms and instructional labs. The department hires undergraduates as teaching, research and computer assistants. There is an active SPS chapter.

– The department, like all at CSM, holds a six-week Summer Field Session devoted to lab techniques and an introduction to research in the department. The intense experience builds community among undergraduate majors.

5. The department recruits new faculty members who match the department's emphasis on undergraduate teaching and research. New faculty members are assigned a senior mentor. Junior faculty members have lunch with the department head almost every day. He evaluates all faculty members every year and provides constant feedback on progress towards meeting the goals of the department. He also holds exit interviews with graduating students in which he seeks evaluation on individual faculty members and courses as well as the program as a whole.

Indicators of Success:

1. The department graduates an average of 22 physics majors per year. The range is 16–28 majors per year and is expected to grow as incoming numbers are growing.

2. About 40% more majors graduate with a degree in physics compared to the number declaring physics majors on entering CSM and 20% of the majors are women.

3. The undergraduate engineering physics program is ABET accredited.

4. Students recognize and appreciate faculty concern for them and the sense of community in the department.

5. The department is known as "friendly," a reputation enjoyed by few other departments.

Keys to Making the Changes:

1. The department has a realistic and very clear sense of its mission that is aligned with the mission of CSM and which is reflected in hiring, in interactions with students, and in interactions with other departments.

2. The department has enjoyed strong leadership from heads who have been able to forge consensus on allocation of resources as the department has shifted its focus from pure teaching to emphasis on teaching and research and to retain emphasis on excellence in undergraduate teaching.

3. The department seeks advice from an outside "Visiting Committee," and from alumni.

4. The university, particularly the Vice President for Academic Affairs, provides resources to support curricular innovation.

5. Colorado School of Mines has an unusually narrow and well-defined mission which carries over to departments and facilitates activities that cross disciplinary lines.

For More Information Contact:

James McNeil, Head, Department of Physics, Colorado School of Mines, Golden, CO 80401; Phone: 303-273-3844; Email: jamcneil@mines.edu.

I. CASE STUDY: Grove City College

The Setting:

Medium-sized undergraduate college with 2,300 students affiliated with the Presbyterian Church. High admission standards and a reputation for offering students a challenging curriculum and keeping fees low. There is no faculty tenure and the college does not accept government funds.

The Department of Physics has five faculty members and no staff although they make heavy use of undergraduate assistants. Forty physics majors in three tracks with 10 graduates per year. Typically, 25% of majors are women.

The majority of graduates successfully seek employment immediately following graduation although a few go to graduate school. Teaching majors are especially sought after.

What Has Been Done:

1. Faculty members work to provide a collegial atmosphere in the department.

    – Faculty work closely together. Offices are clustered to promote informal interaction, and they meet for lunch once a week to discuss teaching and research.

    – Faculty know students well and join them in social events some held in faculty homes and some including alumni.

    – The department holds monthly meals for faculty and students at which students report on research done for example during a summer internship.

    – The department uses many student assistants both as T.A.s in labs and in other capacities so they function as junior members of the department.

2. The department designed and instituted a curriculum that serves the needs of Grove City students. It offers three tracks that lead to jobs or graduate school after graduation.

    – *The Applied Physics Track*: closest to traditional physics major but emphasis is on preparing students for the workplace as well.

    – *The Applied Physics/Computer Track:* replaces some of the advanced physics requirements with courses in computer engineering. Graduates are more attractive to employers than students with only computer skills.

    – *The Physics/General Science Secondary Education Track:* prepares students for careers in secondary teaching. Graduates frequently have multiple job offers on graduation. Physics requirements are kept lean to allow time for better preparation in other sciences and education.

3. The department does an excellent job in teaching service courses using results of PER as well as new technologies. Although physics is no longer required of all students, nonmajors continue to elect physics or astronomy as one of their science options.

    – The department is a campus leader in the use of computers in education. All Grove City students are provided with laptop computers and use them extensively in labs.

    – Eleven years ago the department introduced a nonmajor service course, Fundamentals of the Universe, which has improved the department's reputation among students, faculty and administrators on campus.

4. The department works closely with the Education Department to provide graduates who are eligible for 7–12 certification in physics and general science.

   – The department uses education-track students to critique lectures in the introductory physics course, help with labs and teach study/review sessions, and even give a lecture in the course. They gain practical teaching experience while assisting the department.

5. The department focuses recruiting efforts on students already planning to attend Grove City because it has proved to be a more efficient use of their time and resources than recruiting high school students.

6. The department encourages students to obtain research experience during summers as interns or as participants in REU programs. They have recently begun to build in-house programs designed for student participation.

Indicators of Success:

1. The numbers of students graduating with bachelor's degrees has increased steadily over the last 10 years.

2. Faculty and students within the department work as colleagues.

3. The Physics Department is highly respected on campus.

4. Enrollments in service courses remain strong although physics is no longer a requirement.

5. Graduates on the education track are heavily recruited by school systems.

Keys to Making the Changes:

1. Faculty members in the department make decisions as a group and share a common vision of the department's mission as service to students not a training ground for future Ph.D.s.

2. Faculty members take the scholarship of teaching very seriously and are familiar with the results and applications of PER.

3. Attention to the service courses won the department an excellent reputation on campus and material support from the administration.

4. The department chair has provided strong leadership in implementing the changes.

5. The administration has recognized the growth in the numbers of majors by allowing the department to hire three new faculty members who are keys to building an in-house research program involving undergraduates.

For More Information Contact:

James Downey, Chairman, Department of Physics, Grove City College, Box 2601, 100 Campus Drive, Grove City, PA 16127; Phone: 724-458-2119; Email: jrdowney@gcc.edu.

J. CASE STUDY: Harvard University

The Setting:

Leading research university generally ranked among the top two or three nationally. Physics Department has 40 full-time faculty members, a Director of the Physics Laboratories, a Director of the High Energy Physics Laboratory, a Head Tutor (Director of Undergraduate Studies), an Instructional Laboratory Associate in Physics and 27 Administrative and Support Staff. The department has outstanding and well-funded research programs and a distinguished graduate program.

The Physics Department graduates 50–60 majors (called concentrators) a year; 40–50% of graduates go to graduate school in physics or a closely related field. The remainder pursue a wide range of careers including medical school, law school or business school as well as immediate employment.

Twenty-five percent of concentrators are women; 5% black, 20% Asian, and 6% Latino.

What Has Been Done:

1. Concentrators are required to take a relatively small number of courses relative to other science concentrations at Harvard.

    – The concentration is flexible, and the physics department gives formal recognition to the connections between physics and other disciplines. For example, a combined physics and chemistry concentration, a program of the Physics Department, draws 10–15 students each year. There are other joint concentrations such as Physics-Mathematics, Physics-Astronomy and Physics-History of Science that are done frequently.

2. The faculty are enthusiastic about both undergraduate students and physics.

    – All faculty teach in the undergraduate program although not every year. They take pride in their teaching and continually develop new materials and courses such as the honors introductory course for very well-prepared students. Service courses and core courses to meet the science requirements of nonscience students are considered important as well as courses for majors.

    – The chair stresses the importance of excellent teaching to all new faculty. Many faculty take advantage the services offered by Harvard's Bok Center for Teaching and Learning to improve their teaching. Graduate student T.A.s take a two-day training session from the center followed by sessions of micro-teaching and video taping.

    – Many undergraduates participate in research through an independent research course that allows up to two semester courses credit for participating in independent research supervised by faculty members. The department supports a number of students to do independent research during the summer. Five to 10 undergraduates work as T.A.s helping with large undergraduate sections.

3. The department establishes a community of physics students.

    – The active SPS chapter organizes a "buddy" system that teams first-year students with upper-division concentrators, produces a booklet of advice for new concentrators, organizes lunches and picnics for students and professors, and sponsors weekly "Cool Physics" sessions where a student talks about research. SPS surveyed physics concentrators and shared results with the department. Its officers meet with the chair and head tutor to discuss issues of importance to undergraduates.

– There are many other opportunities for faculty-student interactions, both formal and informal. These activities include study nights, lunches, dinners, weekly presentations by faculty of their research, the Physics Answer Center organized by the graduate students, and undergraduate participation in an annual "puppet show" where graduate students "roast" the faculty members. The physics undergraduates consider themselves active members of a lively physics community.

4. The department insures that all students receive careful advising.

– The head tutor meets individually with all students choosing to enter the concentration to discuss their interests and course plans. She then assigns each student to a member of the faculty who will act as mentor/academic advisor throughout the student's career in the department. The head tutor also remains available for any student needing advice. She checks to see that students complete requirements and are aware of other opportunities.

Indicators of Success:

1. Harvard is one of the leading producers of physics graduates at the bachelor level in the nation.

2. Undergraduates clearly take pride in belonging to a lively, close-knit community. SPS events draw both concentrators and friends of concentrators so that mailings go to 500 students.

3. Three physics faculty have recently won Harvard's Levinson Teaching Prize.

Keys to Making the Changes:

1. The department has a good sense of the capabilities and aspirations of Harvard students and designs a challenging program to meet their expectations. At the same time, the department recognizes that students often have wide-ranging academic interests and career goals and has a concentrators program that is unusually flexible in course requirements with a number of varied options.

2. The department fosters excellence in undergraduate teaching by means of direct comments from the chair, the availability of services of the Bok Center for Teaching and Learning, and department-wide discussions of the undergraduate program at faculty meetings.

3. The department encourages and supports a number of informal interactions among faculty and undergraduates. The culture of the department is that of a lively intellectual community of which undergraduates are important members.

4. The department chairs have played a leadership role in re-enforcing the importance of good undergraduate teaching and keeping a focus on evaluating and rethinking the undergraduate program.

5. The department supports the activities of a lively SPS chapter and uses it to encourage undergraduate input into its undergraduate program.

For More Information Contact:

Gerald Gabrielse, Chair, Physics Department, Harvard University, 17 Oxford St., Cambridge, MA 02138; Phone: 617-495-4381; Email: gabrielse@physics.harvard.edu.

K. CASE STUDY: University of Illinois

The Setting:

A major research department that brings in around $17 million annually in external research support and is nationally ranked in several subfields. Sixty-four tenure-line faculty; 38 affiliate faculty; 59 postdocs.

There were 257 undergraduate majors and 238 Ph.D. students in the fall of 2001. Seventy percent of bachelors-level graduates attend graduate school in some field of physics. The department graduates 25–30 majors per year.

Total undergraduate course enrollment is approximately 2,500 students per semester.

What Has Been Done:

1. A major revision of the introductory sequence in both algebra and calculus-based physics featuring:

    – New labs based on "predict, observe, explain" model

    – Two-hour discussion sections during which T.A.s act as coaches to students working in groups on solving problems

    – Lectures based on power point slides using the department's traditional strength in demonstrations

    – Homework done on the web now includes on-line quizzes and new interactive examples to teach students problem-solving skills

    – In the calculus-based sequence, the former sequence of three 4-credit-hour courses replaced by a more flexible format of two 4-credit-hour courses (mechanics and E&M) plus two 2-credit-hour courses (thermal physics, and waves and quantum mechanics)

2. Development of an infrastructure to support and sustain the revised courses.

    – Introductory courses are taught by teams of faculty who fill roles as lecturers to large (300 student) sections, laboratory coordinator and discussion section coordinator. This has changed the intro courses from isolated, draining faculty assignments into desirable assignments where new faculty can be mentored by experienced team members. Common exams are developed by the entire team.

    – An Associate Head for Undergraduate Programs has been added to the department and the undergraduate staff has been increased to two people who handle all record-keeping and the majority of student problems freeing faculty time for substantive teaching and research.

    – An intense T.A. training program has been introduced to mentor T.A.s in all aspects of teaching but particularly in the interactive methods used in the discussion sections

3. Addition of a staff person with a master's degree in chemistry to manage recruiting, the REU program, the Saturday Physics Honors Program which presents topics in physics to prospective students and the public, and to work with the departmental website. She counsels students who seek to avoid enrollment caps in engineering disciplines by declaring physics majors without any intention of studying physics.

4. A one-credit-hour called Physics Orientation taught by the department head to introduce freshmen to departmental facilities, research programs and career opportunities in physics.

5. Two new courses targeted at definite groups of students: Physics 100 prepares students with weak backgrounds for the calculus-based physics course. Physics 212 is a one-credit "companion" course to the E&M course in the introductory sequence aimed at the most able students to challenge them with extra topics such as relativity and to prepare them for upper-division physics courses.

6. A dozen or so majors participate in an active Physics Society which organizes talks and field trips. Approximately the same number man the Physics Van which presents programs at local schools and museums.

7. A new course beginning in the second semester of junior year and including a summer of research and the first semester of senior year to provide students with research and presentation skills. It is targeted at students who plan to go to graduate school.

8. An active REU program and an astrophysics program involving undergraduates in research beginning in their sophomore year.

Indicators of Success:

1. Greater satisfaction with the introductory physics sequence among client disciplines and students from these disciplines as well as physics majors.

2. Faculty enjoy teaching the introductory sequence as members of a team, and the assignment is no longer considered a detriment to research. New faculty are mentored as members of the team.

3. Enrollments have risen from 57 freshmen physics majors in 1998 to 98 freshmen in 2001. This dramatic rise is not yet reflected in graduation rates which have increased only modestly but should eventually reflect the increase in freshmen.

4. Failure rate in the calculus-based sequence for students who have completed Physics 100 has been halved.

5. Several department faculty have received major university awards for their teaching, and their efforts are recognized within the department's reward system.

6. Student ratings of T.A.s have increased steadily from the beginning of the reform effort. In 1997, 20% of the physics T.A.s were ranked as excellent by their students. In 2001, 75% of the physics T.A.s were ranked as excellent by their students.

7. An active program in Physics Education Research has received external funding and begun to attract graduate students. The work of this research group supports changes in other physics courses.

Keys to Making the Changes:

1. The departmental leadership and most faculty believe that undergraduate education is important to the department. While recognizing its importance, they are pragmatists in recognizing the faculty, particularly junior faculty, must place priority on establishing their research. Thus they have attempted to make changes that allow faculty to do excellent undergraduate teaching without demanding an exorbitant time commitment.

2. The department is located in the College of Engineering which shares its emphasis on the importance of excellent undergraduate education. Excellent teaching is important in promotion and tenure decisions at both the college and departmental levels.

3. Majors are counted at the college rather than the departmental level so the department is not under immediate pressure to increase the numbers of majors. Reforms were implemented only after a year of discussion among faculty and a lengthy process of building consensus.

4. The department has built on a tradition of collegiality as it goes about making changes. The revised courses belong to the team teaching them and to the department as a whole rather than to any single faculty member.

5. The changes in the department have used skilled professional staff to undertake many important tasks such as recruiting which frees faculty members to do the research critical for their success.

6. Because it is large, the department can find internal resources to support instructional improvements by such measures as providing release time to faculty or reallocating staff slots.

7. The improvement of the undergraduate program is a priority for the current department head and his predecessor. They have worked to shift resources to support changes and to establish rewards for faculty who are excellent teachers. The revision of the introductory sequence required substantial released time for faculty and the hiring of support staff.

For More Information Contact:

Gary Gladding, Associate Head for Undergraduate Programs, Department of Physics, University of Illinois, Loomis Laboratory of Physics, 1110 W. Green St., Urbana, IL 61801-3080; Phone: 217-333-0864; Email: geg@uiuc.edu.

L. CASE STUDY: North Carolina State University

The Setting:

Large research-oriented department with close ties to the School of Engineering. The Department of Physics split from the Nuclear Engineering Department in 1963 and about 60% of the department's research effort is applied. There are 36 tenure-line faculty members, 10 research faculty and visiting lecturers, 15 other Ph.D.s, and 15 staff members.

There were 120 undergraduate majors and 85 graduate students in the spring of 2001. Between 1998 and 2001, about one-fourth of physics bachelors graduates enter graduate school in physics; one-fifth continue their education in another field; a fourth find jobs in technical fields; and the remainder pursue other paths. The department graduates about 20 majors per year.

The total undergraduate course enrollment is approximately 2,550 students per semester.

Students are admitted to departments rather than to the university. In 1996, the College of Physical and Mathematical Sciences where physics resides substantially raised admission standards above those of the School of Engineering to stop the influx of students seeking to enter engineering through science departments. College enrollments immediately dropped by one-third, but have since recovered.

What Has Been Done:

1. After in-depth discussion, the department decided to maintain a rigorous traditional physics major. The department is pleased with the success of its traditional B.S. graduates, so has chosen to address its strategic goal of a 50% increase in graduates (to about 30/yr) by developing a new B.A. track, rather than by modifying the B.S. The department gives very high priority to recruiting extremely able students.

2. The department works extremely effectively to mentor undergraduate majors and build community within the department.

    – Majors enter a special section of the introductory calculus-based course with special laboratories and a unique curriculum.

    – New majors are welcomed to the department with a reception early in the academic year and encouraged to join a very active SPS chapter which meets every other week and does four to five activities per semester.

    – Instructors in the introductory course for majors assign group projects to encourage students to work together. By junior year, most physics majors work in informal study groups.

    – A small group of faculty advisors work closely with each class of majors and follows them from freshman year to graduation.

    – There is an undergraduate study room and generous resources for worthwhile projects or that essential supply, pizza.

    – Undergraduates are hired early to work for the department, for example setting up demonstrations or as tutors, in order to involve them in the life of the department.

3. All majors are encouraged to participate in research. The department has a long-standing REU program and presents cash awards for outstanding student research. The majority of undergraduates are mentored by a few faculty members whose research

programs divide into small projects, but there is broad participation by all faculty in mentoring undergraduate research.

4. The senior laboratory has been revised to focus on individual projects that the student must design, review the literature, write up and present to the class.

5. The undergraduate director of the department works closely with the director of admissions of the College of Physical and Mathematical Sciences to attract able students.

    – The undergraduate director calls prospective students, meets with them and their families, and encourages students who wish to transfer into physics majors.

    – The introductory course has an honors section for able engineering students from which the department actively recruits physics majors.

    – The department recruits high school students through mentoring programs involving lab experiences and visits to schools. Other outreach to K–12 students and the public is handled through Science House.

    – Students who do exceptionally well in engineering physics sections receive a letter from their instructor congratulating them on their performance and suggesting that they consider a physics minor. A few such students wind up adding or changing to physics.

6. Students are encouraged to carry double majors, and the department has worked out arrangements with engineering departments to facilitate this by such means as substituting engineering courses for physics requirements, and vice versa.

7. Four years ago, the department established a B.A. degree with reduced requirements in traditional physics to facilitate double majors and attract more students. No new courses were developed to support it. As of May 2001, 17 students have graduated from the program (out of a total of 52 since the first B.A.s were awarded). They have been competitive in the job market, a pleasant surprise to the department head.

8. The department encourages recent graduates who have entered industry to interact with their current students through a banquet for alumni and students. They plan other activities to raise students' awareness of industrial careers that will make use of their large number of successful graduates.

9. The structure of the introductory courses is designed to free faculty for research. The 15 sections of introductory calculus-based physics are coordinated by a single faculty member who makes up the syllabus indicating what topics are to be covered each day, constructs the exams, and places homework assignments on WebAssign which was developed at the university.

10. Three sections of the intro course are offered in the interactive Scale-Up format.

11. There is a thriving and active Physics Tutorial Center.

12. An extensive system of lecture demonstrations was set up by a member of PIRA and is overseen by a staff of students.

Indicators of Success:

1. Very high morale among both faculty and physics majors. They are extremely proud of one another. There is no problem making upper division courses meet enrollment targets, and they expect to replace retiring faculty members.

2. Most faculty and students know one another by name. They form a close-knit community and celebrate others' achievements.

3. Enrollments in physics have returned to their pre-1996 levels despite higher admissions standards.

4. Half the students carry double majors with math in first place and engineering in second.

5. The department has produced three finalists for the Apker Award and one winner.

6. The department has a thriving Physics Education Research Group that has developed the Scale-Up model for introducing interactive instruction to large sections of introductory physics. The group attracts very able graduate students and substantial external funding.

Keys to Making the Changes:

1. The driving force for making changes in the undergraduate physics program is the nearly universal conviction of the faculty that it is important to do an excellent job in preparing undergraduate physics major, not pressure from outside the department.

2. The faculty as a whole participated in the discussions that led to the strategy of concentrating physics on the most able students. There is nearly complete consensus within the department that physics majors are "found" and "authorized" to study physics.

3. The department has a tradition of working collaboratively with engineering departments that has facilitated the development of double majors and other strategies linking departments to the benefit of students.

4. The department faculty work collegially among themselves and with students. The warmth of the faculty for students is a terrifically appealing feature of this department.

5. Department chairs have provided strong leadership in recognizing the importance of the undergraduate program and in providing resources and faculty rewards needed to do it well.

For More Information Contact:

Steve Reynolds, Professor of Physics and Undergraduate Program Coordinator, Physics Department, Box 8202, North Carolina State University, Raleigh, NC 27695-8202; Phone: 919-515-7751; Email: stephen_reynolds@ncsu.edu.

M. CASE STUDY: North Park University

The Setting:

Small, church-related university with 2,300 students that has offered a four-year curriculum since 1958. About 275 incoming freshmen in each class. Twenty physics majors at all levels and an average of five graduating each year.

About 50% of physics majors continue their education after graduation but not necessarily in physics, 14% enter K–12 teaching, and the remainder find employment in a wide variety of positions.

The Physics Department has two faculty members, no staff, and an annual budget for supplies and equipment raised this year to $10,000 from $2,500. Their space is limited to one large lecture/lab area, a laboratory and two offices.

What Has Been Done:

1. The department chair aggressively recruits students who are interested in physics, engineering or science in general and are considering North Park University.

   – She writes (emails) all prospective students interested in physics, engineering or science describing the department and inviting them for a visit.

   – Students visiting campus and their parents visit with her in the department and receive a humorous post card as a follow up.

   – The department chair teaches the introductory physics course so students begin their career already acquainted with one of their professors.

2. The department creates a "home" for physics majors within the department.

   – The lecture/lab room is equipped with a combination lock so physics majors have access 24 hours a day. The room is equipped with computers and has become both an academic and social center for the majors. They remain in the room during the introductory physics course so all majors know all other majors.

   – The departmental atmosphere is kept informal. Faculty and students are on a first name basis with students visiting faculty homes and all members of the department familiar with one another's personal concerns and triumphs.

3. The Physics Department has a very broad definition of success for its graduates. They consider physics an excellent basis for almost any career and teach accordingly.

   – Students are encouraged to carry double majors not only in math and engineering but in areas such as philosophy and languages.

   – Faculty team teach courses with faculty from other disciplines and work closely with colleagues from other academic disciplines.

4. The two faculty offer an academically solid physics major that provides minimum essential preparation for those students who wish to enter graduate school.

   – Majors take a special one-credit hour course along with the introductory sequence to improve their math and computer skills and prepare them for upper-division courses.

   – Introductory labs are taught without formal instructions so students learn to design experiments, prepare proposals, and present their results as 10-minute papers. Group work is introduced in the lecture and homework is submitted via WebAssign.

— Upper-division courses are tailored to the needs of individual students and feature projects, both individual and group.

5. Students are encouraged to participate in REUs and internships off campus in the summers because there is little opportunity for research experiences on campus. The faculty use a network of alums and professional acquaintances to help place students. They collaborate with colleagues at other larger institutions.

Indicators of Success:

1. This very small department consistently graduates an average of more than five majors each year.

2. Morale is very high within the department, and the department is widely respected on campus.

3. The recruiting effort's effectiveness is demonstrated by a decline in majors following the year when the chair was on sabbatical.

4. Students succeed in graduate programs as well as in a variety of other careers.

5. One of the faculty members is very well known and active in the larger physics education community. Both faculty members are intellectually active in their own research areas and successful in obtaining outside funding (not a tradition at North Park).

Keys to Making the Changes:

1. Collegiality and shared vision are absolutely essential if a small department is to build and maintain a strong undergraduate physics program.

2. Change required support from the chair of the Mathematics and Science Division and faculty from other disciplines.

3. Faculty success in obtaining outside funding enabled the purchase of high-quality laboratory and computer equipment and some employment of students on projects.

4. Networking and collaboration with other departments helps the North Park faculty stay intellectually alive.

5. Both faculty members remain intellectually active and excited about physics, teaching physics and their students.

For More Information Contact:

Linda McDonald, Physics Department, North Park University,
3225 W. Foster Ave., Chicago, IL 60625-4895; Phone: 773-244-5665;
Email: llm@northpark.edu.

N. CASE STUDY: Oregon State University

The Setting:

A state-assisted university with about 18,000 students of whom 14,500 are undergraduates. The university has traditional strength in the sciences, engineering, oceanography, forestry, and agriculture. It has been growing steadily in spite of continuing state budget crises.

The Department of Physics has 15 tenure-line faculty and one or two full-time adjunct faculty members recruited to teach the large introductory courses. It has one full-time technician, a second technician shared with chemistry, and four office staff who also administer departmental grants. The department currently enrolls about 35 Ph.D. candidates and attracts about $1.5 million annually in research funding including several large grants aimed at undergraduate education, and it is in the process of developing a professional masters program with funding from Sloan.

The department offers B.S. degrees in Physics and Engineering Physics and initiated a new B.S. degree in Computational Physics this year. The department currently enrolls about 110 undergraduate majors in all years.

The Department of Physics receives an unusually high percentage of its majors as transfer students from two-year colleges or other institutions. About a quarter of undergraduate majors transfer in for the sophomore year and another quarter enter before their junior year. A quarter declare physics majors as entering freshmen and another quarter declare majors in engineering physics although many of them eventually become physics majors.

What Has Been Done:

1. In 1996, the department undertook a major revision of its upper-division physics curriculum to allow flexibility in scheduling so that engineering physics majors could participate in an internship and to help students make the transition in difficulty from lower-division physics courses to upper-division courses.

   – The Paradigms Project reorganized the junior-level curriculum into nine 3-week-long paradigms courses, each of which focuses on a class of physics problem that appears in more than one subdiscipline of physics, oscillations or static vector fields for example. In addition, the Paradigms courses introduce new, research-based pedagogies that encourage students to work and study in groups and to take responsibility for their own learning.

   – In their senior years, students take capstone courses in traditional subdisciplines of physics. Because students have been exposed to some material in the paradigms courses, the capstone courses can cover material more rapidly. They are taught more traditionally than the paradigms to prepare students for graduate school experiences.

2. The department has built on its traditional strength in computational physics to introduce a new undergraduate major in the field. The program has received external funding from NSF and DOE. The first students are enrolling now.

3. The Department of Physics was recently selected as one of six primary sites for the Physics Teacher Education Coalition. This program will introduce two new majors appropriate to preservice teachers as well as promote revision in the calculus-based introductory course.

4. The department recruits physics majors not only by supporting the recruiting activities of the College of Sciences, by responding to inquiries from prospective students, and by maintaining collegial relations with local high school teachers and TYC faculty, but also by actively recruiting talented students in the modern physics course which is taken by students from several engineering disciplines.

5. Students form a community, particularly after the paradigms sequence during which they form close-knit study groups. They have access to a study room with all amenities including tables with white board tops on which they can write, access to the paradigms classroom and to a computer room. The department has Sigma Pi Sigma and SPS chapters.

6. Students are assigned to a faculty advisor as soon as they declare a physics major. The Director of Undergraduate Programs oversees advising and is known as the person to come to for answers to scheduling and other problems. Faculty are accessible to students, particularly during the paradigms courses when they work intensely with the students in their courses.

7. All physics majors are required to complete a research project. Most faculty advise an undergraduate or two on research projects although some of them work with advisors in other science departments or meet the requirement through participation in REU programs.

8. Revisions introduced into the algebra-based include lectures with an electronic response system, concept-based laboratories, and the introduction of tutorials in recitation sections.

Indicators of Success:

1. The number of physics graduates has increased and continues to increase. In 1999, 12 majors graduated; in 2000, 15 majors graduated, and in 2001, 22 majors graduated.

2. Since the introduction of the Paradigms and Capstone courses, GRE scores have held steady, even though more and weaker students take the exam. In addition, students have formed a much closer community and rate the paradigms courses as the best part of their undergraduate physics experience.

3. Nearly all faculty members are actively involved in one of the undergraduate projects.

4. The administration recognizes and respects the work of the Department of Physics.

5. Faculty from client disciplines respect the work of the physics department in education and, even when there are problems, express confidence that the department will fix them.

Keys to Making the Changes:

1. Major revisions are supported by a strong faculty consensus, and all faculty take responsibility for the excellence of the undergraduate program in physics.

2. Department chairs have provided strong leadership in recognizing the importance of the undergraduate program and building faculty support for revisions.

3. External funding, particularly from the National Science Foundation, has played a key role in making major curricular revisions possible during a time of contracting state budgets.

For More Information Contact:

Henri Jansen, Chair, Department of Physics, Oregon State University,
301 Weniger Hall, Corvallis, OR 97331; Phone: 541-737-4631;
Email: chair@physics.orst.edu.

O. CASE STUDY: Reed College

The Setting:

An independent and highly selective private liberal arts college enrolling about 1,400 students. The college offers 22 department-based majors and 12 interdisciplinary majors. Students must complete a core curriculum in arts and sciences. All students must pass a qualifying exam in the junior year, and all must complete and defend a senior thesis.

Reed maintains a casual atmosphere between faculty and students; students and faculty are all on a first-name basis. Grades are de-emphasized; professors assign grades in each course, but students do not routinely receive a grade report.

The physics department consists of six tenure-track faculty and one adjunct. Support staff include a secretary, a machinist, and an electronics technician.

Reed awards an average of 19 physics degrees each year, which represents about 7% of the total number of baccalaureate degrees awarded by the college and about 30% of the math and science degrees. About 20% of the physics degrees are awarded to women. Typically 50% of the graduates enter graduate school in physics or related areas. Very few become high-school teachers.

What Has Been Done:

1. The college offers a traditional undergraduate curriculum leading to the B.S., but students can also earn degrees in one of three alternate tracks: a joint chemistry/physics program, a joint mathematics/physics program, and a 3/2 engineering program (in which the final two years of engineering work is taken at Columbia or Cal Tech). About one-third of the degrees are earned in the alternate tracks.

2. Two years of introductory physics are offered. The second-year course, which enrolls about 30 students (mostly physics majors), is taught in the lecture/laboratory/conference mode. The laboratory has been modified to emphasize acquiring electronics skills and learning to solve problems using Mathematica. Individual faculty-student conferences are used to critique the lab work.

3. A major effort has gone into developing the junior-year laboratory experience. The first semester emphasizes electronics, and the second semester is based around student projects using sophisticated apparatus to perform contemporary experiments. The course makes extensive use of LabVIEW, which permits experiments to be computer interfaced and controlled. These course revisions were supported by grants from the NSF and the Murdock Trust.

4. The department has recently received a $500,000 Keck Foundation grant (matched by the college) to undertake a revision of the instructional laboratory program, which will include developing inquiry-based labs for the first-year course and upgrades of the equipment for the second-year course and the advanced lab. The grant will fund a lab manager to support the course improvements; the college will continue to support that position when the grant is concluded.

5. The senior thesis is a unifying element that forges bonds among the students as well as between students and faculty.

Indicators of Success:

1. Reed ranks among the top small colleges in the United States in number of physics degrees awarded and exceeds 80% of the Ph.D.-granting institutions. The number of physics graduates has remained steady or increased while the national total has declined.

2. Retention of majors is very high.

3. There is a cohesive, supportive, and collegial attitude that pervades the department. Students are mutually supportive and do not compete with one another for grades

4. The department has received substantial extramural funding for course and curriculum development.

5. The department is well respected on campus and has the strong support of the college administration.

Keys to Making the Changes:

1. The faculty is committed to curricular reform.

2. The department has deliberately chosen to emphasize undergraduate education over funded research as a measure of faculty accomplishment, and the college administration has supported that decision (even though other science departments at Reed take a more traditional view of the faculty reward system).

3. External funding and college matching funds have enabled laboratory course improvements.

4. The conferences in the second-year course serve to develop the close bond that forms between faculty and students.

5. Students are invited to serve as undergraduate T.A.s as early as the sophomore year. This attracts students into the major and builds the cohesive spirit in the department.

6. The senior thesis is a unifying element and a capstone experience for the students.

7. The department maintains an active seminar program that involves undergraduates.

8. The college and the department set high standards and then focus their resources to assist students in meeting those standards.

For More Information Contact:

John Essick, Chair of the Physics Department, Reed College,
3203 SE Woodstock Blvd., Portland, OR 97202; Phone: 503-771-1112;
Email: jessick@reed.edu.

P. CASE STUDY: Rutgers University

The Setting:

A public, land-grant, Research 1 institution with about 26,000 undergraduate students. The university includes several colleges, in particular Rutgers College which awards most of the undergraduate physics degrees, Douglas College, a woman's college, and a School of Engineering.

The Department of Physics and Astronomy consists of 65 faculty members, of whom 20% are astronomers, four are research professors, and several have joint appointments with various research centers. There are five full-time instructional support staff members.

Women represent about 20% of majors, Asian 13%, Hispanics 3% and 5% are of mixed race. 20% of majors received a 4 or a 5 on one of the Physics AP tests, and a quarter of majors are transfer students, often from New Jersey community or county colleges. Twenty to 25% of students go to graduate school usually in physics or astrophysics, one to two become K–12 teachers, 10% go to professional school, and the remainder enter the workforce, commonly in technical jobs.

What Has Been Done:

1. Undergraduate advising has been centralized with one capable and caring advisor working with all 150 undergraduate majors.

    – Advice is consistent and professional and students feel that the advisor and hence the department cares for their success. He maintains an open door policy for students.

2. The department offers four options for a physics major.

    – The Professional Option in Physics or Astrophysics (B.S.) leads to a B.S. and prepares students for graduate school. The General Option (B.A.) is designed for students who are interested in physics but do not intend to become professional physicists. The Dual Degree option allows students to obtain an engineering degree and a degree in physics.

3   The department offers a variety of introductory courses in physics that respond to the diverse academic preparation of entering students.

    – A three-semester honors sequence targets well-prepared students; a two-semester course serves students in the General Option and biological science students; and a four-semester sequence is taken by engineering majors and physics majors not enrolled in the honors sequence. "Extended" versions of the first two semesters of the four-semester sequence and the two-semester sequence are offered for one extra credit per semester and target students who are less well-prepared. The "extended" courses use cooperative workshops and integrated labs to provide extra time for students to absorb physics. In addition, there is an experimental algebra-trig course in place at Douglass College featuring group workshops and an integrated laboratory.

    – The department created a Physics Learning Center which was generalized and now operates as the Math and Science Learning Center to provide review sessions, physics videotapes, copies of old exams, and a congenial place for students to meet and study.

4. The department has revised introductory laboratories to emphasize more of a discovery experience and to make effective use of computers. The primary physics lecture hall has been fitted with a state-of-the-art audio visual system and a personal response system, a staff member works to develop and prepare demonstrations, and half a dozen courses use WebAssign.

5. Students are encouraged to participate in research through taking jobs in faculty labs or through participation in the department's Honors Program which requires a senior thesis. Honors students' work is showcased in a variety of department forums.

6. The active SPS chapter runs a free, informal tutoring service for students in the introductory courses. Students have the use of a room in the Math and Science Learning Center for two nights each week.

7. The department has an active T.A. training program including visiting and video-taping each T.A.

8. The department participates in university recruiting activities such as the Open House for high school students as well as providing personal tours for interested students and their parents. In addition, several faculty members are involved in projects that introduce students and teachers to new developments in particle physics and astrophysics as well as in demonstration shows for school groups and the general public.

Indicators of Success:

1. The department currently graduates about 40 students per year which represents a doubling in the number of majors since 1980.

2. Approximately 0.6% of the undergraduate degrees awarded by Rutgers are in physics, more than double the national average for physics departments.

3. Students speak warmly of the sense of community in the department and faculty efforts to involve them in research. They feel the department compares very favorably to other science departments at Rutgers.

4. Students in "extended" introductory courses obtain exam scores that duplicate those of students in the regular sequence.

Keys to Making the Changes:

1. The department appointed an Associate Chair who serves as Undergraduate Program Director and takes full responsibility for all aspects of the program and serves as advisor to all undergraduates.

2. The department has aggressively sought and obtained university and NSF funding to support improvements in instructional facilities and equipment.

3. A generous donation allows the department to offer seven full-tuition scholarships for majors each year as well as two summer research internships.

> **For More Information Contact:**
>
> Mohan Kalelkar, Undergraduate Program Director, Department
> of Physics and Astronomy, Rutgers University, P.O. Box 849,
> Piscataway, NJ 08855-0849; Phone: 732-445-3878;
> Email: lelkar@physics.rutgers.edu.

Q. CASE STUDY: State University of New York at Geneseo

The Setting:

A highly selective public four-year college with a total student population of about 5,100 students. Geneseo prides itself on being the best liberal arts college in the SUNY system, and in addition has Schools of Business, Education and Performing Arts as well as several masters programs. The college is increasingly emphasizing research, publication, and creative accomplishments.

The Department of Physics and Astronomy has eight tenure-track faculty members, of whom one is on research leave and has been replaced by a visiting assistant professor and a second serves as Associate Provost. Six of the tenure-track faculty members have been hired since 1993. Staff consist of a senior instructional support specialist and a full-time secretary. The department employs junior and senior physics majors as undergraduate lab instructors who work with faculty in the labs.

The department offers a B.A. degree in physics, a B.S. degree in applied physics and participates in a 3/2 engineering program. About 17% of physics graduates pursue Ph.D. programs in physics; 28% enter Ph.D. or masters programs in engineering; 13% pursue graduate or professional degrees in other areas; 14% enter the workforce; 11% complete 3/2 engineering programs, and 6% enter public school teaching.

What Has Been Done:

1. Following a tradition built at its founding in the 1960s, the department emphasizes close student-faculty interactions.

    – Faculty maintain an open-door policy for students. The department promotes an active physics club and an annual bridge-building contest to build departmental community. Faculty hiring efforts are aimed at bringing in faculty who will maintain that tradition.

    – Undergraduate majors get keys to the building and a computer area, and they use instructional labs in the late afternoon and evening. Faculty offices are close to the student areas, and students find they can talk to faculty about anything, including courses faculty are not currently teaching.

    – The department keeps photographs of all majors and faculty members on bulletin boards near the department office. They maintain a map showing the location of previous graduates. The department uses its weekly colloquium for a mix of research talks, talks about careers and student activities such as the bridge building contest or Physics Bowl.

2. Students who enter Geneseo with an expressed interest in physics are encouraged to enroll in a one-credit First-Year Physics Experience course that allows students to meet one another, upper classmen and the physics faculty.

3. The department recruits students when they are first admitted to Geneseo. They contact all students who express an interest in physics and astronomy as well as those whose academic preparation suggests they can succeed in physics. They actively recruit physics majors in the introductory course. About 40 students a year express interest in the 3/2 engineering program, and many of them convert to physics majors.

4. The B.A. in physics has minimum requirements allowing flexibility. Most students take more physics than the minimum requirement, and the B.S. in applied physics requires more courses including electronics and computer science.

5. Well over half the physics majors have a summer research experience either at Geneseo or in an off-campus research program.

    – Six of the current faculty members have externally funded research projects. The department uses its two MeV Van de Graff accelerator to calibrate detectors for laser fusion efforts at the University of Rochester and Lawrence Livermore National Laboratory.

6. The department works to extend the feeling of community to students in service courses.

    – Faculty take pictures of all students during the first week of classes and learn all names within a week or so.

Indicators of Success:

1. In the last nine years, the department has graduated an average of 17 physics majors per year of whom 22% were women and 4%, minorities. There are currently 76 declared majors in all four years.

2. From their freshman years on, students express a strong sense of belonging to the physics department. They use the word "family" to describe their experience. They are aware of anecdotes about past students and events.

3. The department has the reputation of being a challenging but friendly department that provides a high quality academic program.

4. The administration holds the department in high regard because of their devotion to teaching and time spent involving students in research.

Keys to Making the Changes:

1. The physics department has a strong and clear sense of its mission that is shared by all faculty members within the department.

2. Communication among faculty members has high priority. Physics faculty members have lunch every week which allows the faculty members to communicate informally. In addition, they have a formal faculty meeting once a week. Generally important decisions require much discussion and are resolved by consensus.

3. The department has had strong leadership from chairs who have been able to form a consensus on the mission and direction of the department.

4. The entire department has worked to recruit faculty members who will support the department's mission and to keep them. The chair chats with them informally but regularly as they become part of the physics "family."

5. The college also works to integrate junior faculty into the larger community through putting them on committees in their second years and through a Teaching/Learning Center.

6. The administration provides strong institutional support for the mission of the department which fits well into the overall mission of Geneseo.

For More Information Contact:
Kurt Fletcher, Chair, Department of Physics and Astronomy, SUNY Geneseo, 1 College Circle, Geneseo, NY 14454; Phone 716-245-5281; Email: fletcher@geneseo.edu.

R. CASE STUDY: University of Virginia

The Setting:

Moderate-size state-supported research university with 12,500 undergraduates and 6,000 graduate and professional students. The Physics Department has active research in the major fields of physics, particularly at Thomas Jefferson National Accelerator Laboratory, and $5 million in external funding each year.

The department has 34 tenure-line faculty members, five research faculty, a general faculty members who is Director of Laboratories, a visiting professor, a lecturer and 26 research associates and scientists. There are 25 supporting staff and 72 graduate students.

On average, 32 physics majors graduate each year. Twenty-five percent were women and 30% were double majors. Twenty-five percent went to graduate school in physics and another 25% continued their studies in another field or in professional school. Two-thousand majors and non-majors enroll in 25 undergraduate courses each semester.

What Has Been Done:

1. In 1995, a new and more flexible B.A. degree designed for students who do not plan to attend graduate school went into effect with reduced requirements in math and upper division physics.

    – Two new courses, "Widely Applied Physics," treat principles of physics from the perspective of modern applications and form a key component of the new degree.

2. In the period 1991 to 1993, the department instituted two new courses designed for liberal arts students. The two-semester "How Things Work" and the one-semester "Galileo and Einstein" courses have attracted large numbers of students to the department helping increase the numbers of student credit hours generated each semester. These courses, as well as all of the introductory courses for majors and non-majors, make extensive use of lecture demonstrations provided by a well-equipped Lecture Demonstration Facility, whose staff of two full-time and a part-time member provide more than 2,000 individual experimental demos each semester.

3. The introductory and upper-level laboratory courses for physics majors have been revitalized and upgraded using nearly $200,000 in internal and external funding. Computers are used for data acquisition and analysis at all levels, and techniques learned in the introductory laboratory are coordinated with practices in the upper-level laboratory courses. The labs incorporate new experiments in modern physics. Most undergraduate physics majors work on research in faculty labs where they are able to make use of techniques learned in the new laboratory courses and to carry out independent study projects.

4. When the number of credit hours required for engineers was reduced from 12 to eight, a joint committee of engineering and physics faculty worked to design new studio-style workshops to replace conventional laboratories and recitation sections.

    – The new workshops are carried out in a physics lab with 24 students per section, working in cooperative groups of three on labs and problem solving. The workshops incorporate the methods of Real Time Physics and use WebAssign. Lectures in the course use interactive lecture demonstrations and WebAssign for problem assignments. A popular resource for this course is a room staffed by a teaching assistant during afternoons and early evenings where individuals and groups of students come to ask questions and receive tutorial assistance.

5. The department created a Master of Arts in Physics Education program designed for high school and middle school teachers who do not have an undergraduate degree in physics. The Physics Department offers 14 courses for the professional development of K–12 teachers to improve their competency in physics and to assist them in obtaining endorsement or recertification. The department offers a course based on *Powerful Ideas for Physical Science* for pre-service elementary and middle school teachers.

6. The process of revitalizing the department is continuing. Current projects include three new courses in computational physics and three new courses in optics to support new optional concentrations in computational physics and optics.

Indicators of Success:

1. Since 1995, the department has graduated an average of 32 physics majors per year. Before 1995, they averaged 20 majors per year.

2. The B.A. has attracted new majors to the department without producing a significant decline in the number of majors in the traditional and rigorous B.S. program.

3. The two new liberal arts courses attract around 900 students each year. "How Things Work" has generated a textbook and national attention and imitation.

4. About two-thirds of faculty are actively involved in undergraduate teaching at the present time.

5. Morale among physics majors is very good, and majors are pleased with their opportunities for informal interaction with the faculty.

6. The Physics Department is viewed very positively by most of the students taking courses in the department, and approximately a third of all University of Virginia students take at least one course in physics before graduating.

Keys to Making the Changes:

1. The department as a whole is very involved with undergraduate education and feels that undergraduate education is an important part of its members' professional responsibilities. Ideas for change come from the faculty and are channeled through an active undergraduate committee.

2. The college administration encourages changes that increase the numbers of majors and the enrollments in service courses. Administrators have both put pressure on the department to change and supported efforts to do so.

3. The Department of Physics has an excellent working relationship with the College of Engineering allowing the formation of joint committees when problems arise in the service courses. The department also maintains good relations with the Department of Astronomy. About five majors per year graduate with a B.A. in Astronomy/Physics.

4. The environment in the department is dynamic and open to change.

5. The department is acutely aware of its role in educating future K–12 teachers and providing inservice activities.

6. The department has succeeded in attracting nearly half a million dollars from internal and external sources to support revision of its courses.

> **For More Information Contact:**
> Bascom Deaver, Associate Chair, Department of Physics, University of Virginia, 382 McCormick Road, Charlottesville, VA 22904; Phone: 804-924-6575; Email: bsd@virginia.edu.

S. CASE STUDY: Whitman College

The Setting:

A private liberal arts college with an undergraduate enrollment of 1,300 students located in the southeast corner of Washington State. An excellent regional college with a faculty strongly committed to teaching and strongly coupled to their students.

The Physics Department has four tenure-line faculty members, one of whom currently teaches half time in the general studies program. The department is authorized to hire a biophysicist but has been unable to fill the slot so that the position is currently filled by a visitor with another specialty. Three of the four tenure-line faculty have been hired since 1990.

The number of majors (10 graduates per year) has remained nearly constant despite the turnover in faculty. About one-third of the students are women. Fifty to sixty percent of graduates attend graduate school and 10–20% enter high school teaching.

What Has Been Done:

1. The faculty cares deeply about students, and the students know and appreciate this concern. Doors are open, and faculty always make time for students with questions about any aspect of physics. Small sections in the introductory physics sequence allow physics majors to become well acquainted with faculty members as freshmen.

2. The department has established 3/2 programs with engineering schools at Washington University at St. Louis, Columbia University, and Cal Tech under which students obtain a B.A. from Whitman and a B.S. from the engineering program. These programs attract majors to Whitman, and many of them elect to remain at Whitman for four years and graduate as physics majors.

3. The department participates in a program of combined majors which allows the physics department to benefit from some of the unique strengths of Whitman including departments of astronomy and geology.

   – The Physics Department has decided that the flexibility inherent in the combined majors compensates for less depth in physics. The departments from which a combined major graduates each receive half credit for the major. Majors at Whitman are sized so that they require about one-third of the credit hours needed to graduate. General studies courses require another third, and the remaining third provides students options to tailor their academic programs to their individual interests.

4. A grant from the Howard Hughes Medical Institute has enabled revision of the introductory physics sequence and provided support for a tenure-line faculty member in biophysics.

   – The introductory sequence consists of two semester-long courses, taught in classes no larger than 25 students with integrated laboratories. The courses stress depth and conceptual understanding rather than encyclopedic coverage. There are usually four sections of each of the introductory courses and they are taken by chemistry, physics, geology and biochemistry majors. However, scheduling sections geared to special student interests has proven difficult.

   – The introductory labs stress measurement and analysis, and the sophomore labs can allow students to design experiments or extend existing ones in order to equip them with skills needed for research.

5. The department has created a one-credit course which meets at lunchtime to hear outside speakers talk about what scientists and engineers actually do. Lunch is provided for students who are not on the meal plan.

6. Faculty are now expected to be active in research, and student participation in that research is encouraged. On average, six students per year spend the summer working in faculty research labs.

Indicators of Success:

1. The Physics Department graduates 10 majors per year, a number in which combined majors like Physics-Mathematics and Physics-Astronomy are counted as 0.5.

2. Several Whitman physics graduates have gone on to extremely distinguished careers in physics.

3. Physics majors appreciate the attention they receive from faculty members and love the introductory course sequence.

4. Students taking the MCAT and who have gone through the introductory physics sequence average at the 80th percentile on the part of the exam that includes both physics and chemistry.

5. Faculty are successful in involving students in scholarly projects and training them for careers in graduate school as well as to directly enter the workplace.

Keys to Making the Changes:

1. The new, younger faculty recently hired have brought enthusiasm and energy to the department's programs in teaching and research.

2. The college provides a program of merit scholarships that help to attract talented students to the physics department among others. Over the last decade, the college has attracted increasingly talented and well-prepared students.

3. Whitman offers a full year of sabbatical leave after four years of full-time teaching.

4. Collaboration with the biology department has provided leadership in seeking the funds that have allowed the reform of the introductory course sequence.

For More Information Contact:

Mark Beck, Chair, Department of Physics, Whitman College,
Walla Walla, WA 99362; Phone: 509-527-5260;
Email: beckmk@whitman.edu.

T. CASE STUDY: The University of Wisconsin–La Crosse

The Setting:

Comprehensive, state-supported university with a total enrollment of 9,200 students of whom 8,500 are undergraduates. Capped admission has made academic preparation of students, largely from Wisconsin and Minnesota, second highest among institutions in the University of Wisconsin system. La Crosse is the center of a metropolitan area of 100,000 on the border of Wisconsin and Minnesota located 140 miles from Madison and 160 miles from Minneapolis.

The Department of Physics has seven tenure-line faculty, including an astronomer who is planetarium director, and two part-time faculty, a three-fourths-time departmental secretary, and a full-time electronics technician.

The department graduates about 15 majors per year. About half graduate from La Crosse after the first year in engineering school on a 3/2 program. The majority of the others go to graduate school (most frequently in optics because of the department's research interests) and the others enter the workplace.

What Has Been Done:

1. The undergraduate physics curriculum has been completely overhauled.

    – The department has 3/2 programs with the University of Wisconsin campuses at Madison, Milwaukee, and Platteville and with the University of Minnesota. Admission for students completing the required three-year curriculum at La Crosse with a high enough grade point average is automatic.

    – The upper-division curriculum introduces specialization early so students graduate in physics or in physics with emphasis in optics, computational physics, astronomy, biomedicine, or business.

    – Students enter the major from the algebra-based introductory sequence as well as the calculus-based sequence. Their math skills even out by junior year, and the system allows all potential majors to immediately begin course work in physics and opens a much larger pool of introductory physics students to the department's recruiting efforts.

    – All laboratories have been reworked and equipped with computers and other modern equipment. Equipment from the sophomore level up is very modern and has the feel of a research lab.

2. All majors are required to participate in research or some appropriate capstone experience such as student teaching and receive course credit for doing so. The college supports this undergraduate research initiative with student research grants, travel grants for students and summer fellowships.

    – Faculty actively recruit students to research projects beginning with their arrival on campus. Many students publish and make presentations at national and regional meetings.

    – Course loads are figured on the basis of contact hours so faculty receive load credit for working with undergraduate research. They are the only department in the university to do this.

3. All freshmen and sophomore physics majors are very strongly encouraged to enroll in a one-credit hour seminar course.

   – This course, attended by 50 plus students and all faculty, is used as a primary means to build community in the department. The department provides information on research opportunities in and out of the department and brings in speakers from industry and the area to talk about using physics as a basis for a variety of careers.

4. The Physics Department aggressively recruits students and works to retain them once they arrive in the department.

   – They target likely physics majors for personal letters, award freshman scholarships, hold open houses for prospective majors and their parents, and present a very popular laser light show for middle school students. The department considers advising a critical element of students' success and works hard to provide each student with contact with a faculty advisor from the beginning. Students cite faculty mentoring and friendship as the best thing about their experience in the department.

5. The department runs summer workshops for in-service teachers and has created a physical science course for pre-service elementary teachers using inquiry-based instruction with no lectures. The pre-service course is so successful that other colleges would like to see it offered for their students even though the department does not have the personnel to do so.

6. The department pays attention to publicizing its programs within the university and the local community.

   – The chair has good working relationships with the local press who provide him with coverage for the department's activities. The department hosts an annual Distinguished Lecture Series in Physics which brings Nobel laureates to campus for the usual lectures in addition to a major banquet for local community and industrial leaders.

7. The faculty, particularly the chair, work to build community within the department.

   – Faculty members' assignments reflect their unique strengths. Junior faculty are mentored by senior faculty towards success in achieving tenure. Hires have been made carefully to strengthen the department and increase its morale.

Indicators of Success:

1. The department graduates 15 majors per year and this number continues to grow. In 1990, the department graduated about one major every other year.

2. The department's efforts are respected and supported by all levels of administration.

3. Two faculty members have won teaching awards in the last few years.

4. Faculty and students are actively publishing and presenting papers at meetings. Many faculty papers have student co-authors.

5. Students leave the program positioned for success in graduate programs and in engineering schools.

6. The department works as a team with a shared sense of mission and a real respect for one another's contributions to the work of the department.

Keys to Making the Changes:

1. The Physics Department has enjoyed sustained administrative support. Revitalization of the department began when the dean brought in an outside chair and another experienced faculty member. The administration invested resources to attract good people and to provide them with the tools they need to make effective changes.

2. The department chair provides very skillful personal leadership to the department. He leads by example, works to build consensus within the department, and enjoys great respect from university and college administrators.

3. The department works hard to build a common sense of mission and to use limited resources and people in the most effect way possible.

4. Curricular revisions have been carefully designed to meet the needs of students and are revised in response to student feedback. The department chair assigns the very best teachers to the introductory courses.

5. The department's increased emphasis on research has not diminished its focus on excellent teaching. All faculty consistently demonstrate a genuine concern for students that is recognized by the students.

6. While working hard, the department maintains a humane atmosphere where families are considered important.

7. The efforts by the department to publicize its programs have paid off in efforts to attract resources and students.

For More Information Contact:

Gubbi Sudhakaran, Chair, Physics Department, University of Wisconsin–La Crosse, La Crosse, WI 54601; Phone: 608-785-8431; Email: sudhakar.gubb@uwlax.edu.

U. CASE STUDY: Lawrence University

The Setting:

A small nondenominational, liberal arts university with about 1,300 students, a quarter of whom are enrolled in the Conservatory of Music but may double major in other disciplines. The department has four tenure-line faculty members, a visiting assistant professor, a part-time electronics technician, and a part-time machinist.

The department graduates an average of 10 physics majors each year on a single degree program although some of them also obtain a bachelor's degree in music. Approximately 50% of the graduates pursue graduate study in physics.

What Has Been Done:

1. The department offers a fairly traditional, extremely rigorous physics major and is proud of its high standards. Faculty expect students to work very hard and treat them as junior colleagues. Faculty maintain an open-door policy for helping students and clearly view undergraduate education as their core mission.

2. The department has developed signature programs in laser and computational physics, and is developing signature programs in surface physics and nonneutral plasmas.

   – A signature program is first a teaching program but has especially well-equipped laboratories and ties to faculty research so that it generates specialty courses, promotes student/faculty interactions, increases departmental pride, and supports student projects. They also help to recruit students.

3. The department works hard to attract to Lawrence talented high school students who are already inclined to major in physics.

   – The department holds an annual weekend workshop for high school seniors with strong interest in physics. The faculty select 26–30 participants from 50 or so applicants to attend a spring workshop with all expenses, including air fare, paid by Lawrence. Each participant is hosted by a Lawrence physics major and spends a day in the signature program laboratories. Approximately 30% of the attendees matriculate.

4. The department involves students in departmental affairs as contributors to curricular discussions, as participants in interviewing candidates for positions and entertaining visitors, and as laboratory assistants and help session leaders for introductory courses.

   – The department holds twice-weekly teas, an annual picnic and an annual weekend retreat. Students have 24/7 access to a student common room, labs for student research and the Computational Laboratory.

   – There is a chapter of the SPS, a Sir Isaac Newton Society, and a Women of Physics club.

5. The faculty strongly encourage student involvement in research by recommending a capstone experience for their majors and encouraging them to spend a summer either at Lawrence working with a faculty member or in a REU program elsewhere.

6. The department has developed several courses to introduce nonmajors to physics, some of which have laboratory components and some of which are oversubscribed.

7. The department currently offers an optional course in Computational Tools in Physics, but students generally lack time to participate in this course. The department is currently changing the second-year mechanics course to include computational methods. In addition, it is developing a signature program in surface physics and enhancing the programs in laser physics and computational physics with a $400,000 grant from the Keck Foundation and an additional $157,000 in matching funds from Lawrence University.

Indicators of Success:

1. The department graduates an average of 10 physics majors per year.

2. Since 1987, the department has received almost $2.5 million in external funding from Research Corporation, NSF, the Sloan Foundation, the Keck Foundation, and others.

3. Morale among students is very high, and students frequently expressed satisfaction with their relationships with faculty and fellow students as well as the preparation they were receiving for future graduate study or careers in physics.

4. Since 1991, GRE scores of graduates have risen appreciably, and many graduates receive awards such as NSF graduate fellowships, Clare Boothe Luce scholarships, a Rhodes scholarship, and a Hertz scholarship.

5. The department retains almost every major attracted into its program.

6. The sense of community that the physics department creates and the excellence of its program are well respected by other science faculty and the administration at Lawrence.

Keys to Making the Changes:

1. In the mid 1980s, two faculty members became dissatisfied with the department's record in graduating an average of five majors per year. These individuals played leading roles in revitalizing the physics program at Lawrence.

2. The president of Lawrence University has strongly supported changes in the physics program. The university has provided nearly one million dollars in matching funds to support the department's development and search for external funds.

3. The Research Corporation has provided the department with two external faculty consultants as well as advice from leaders of the Research Corporation to assist the president of the university and the department in making changes.

4. Students are expected to work hard in physics, but their interests are protected as members of the physics family. For example, faculty offices are smaller than offices in the other sciences so that there is room for a student study room/lounge in the corridor with the faculty offices.

5. Faculty view undergraduate education as their core mission and emphasize the view that an undergraduate physics program is much more than curriculum.

For More Information Contact:

David M. Cook, Department of Physics, Lawrence University,
Box 599, Appleton, WI 54912; Phone: 920-832-6721;
E-mail: david.m.cook@lawrence.edu.